Ecological Studies

Analysis and Synthesis

Edited by

W.D. Billings, Durham (USA) F. Golley, Athens (USA)
O.L. Lange, Würzburg (FRG) J.S. Olson, Oak Ridge (USA)
H. Remmert, Marburg (FRG)

Volume 72

Ecological Studies

Volume 59
Ecology of Biological Invasions of North America and Hawaii
Edited by H.A. Mooney and J.A. Drake
1986. X, 320p., 25 figures. cloth
ISBN 0-387-96289-1

Volume 60
Amazonian Rain Forests Ecosystem Disturbance and Recovery
Edited by Carl F. Jordan
1987. X, 133p., 55 figures. cloth
ISBN 0-387-96397-9

Volume 61
Potentials and Limitations of Ecosystem Analysis
Edited by E.-D. Schulze and H. Zwolfer
1987. XII, 435p., 141 figures. cloth
ISBN 0-387-17138-X

Volume 62
Frost Survival of Plants
By A. Sakai and W. Larcher
1987. XII, 321p., 200 figures, 78 tables. cloth
ISBN 0-387-17332-3

Volume 63
Long-Term Forest Dynamics of the Temperate Zone
By Paul A. Delcourt and Hazel R. Delcourt
1987. XIV, 450p., 90 figures, 333 maps. cloth
ISBN 0-387-96495-9

Volume 64
Landscape Heterogeneity and Disturbance
Edited by Monica Goigel Turner
1987. XII, 241p., 56 figures. cloth
ISBN 0-387-96497-5

Volume 65
Community Ecology of Sea Otters
Edited by G.R. van Blaricom and J.A. Estes
1987. X, 280p., 71 figures. cloth
ISBN 3-540-18090-7

Volume 66
Forest Hydrology and Ecology at Coweeta
Edited by W.T. Swank and D.A. Crossley, Jr.
1987. XIV, 512p., 151 figures. cloth
ISBN 0-387-96547-5

Volume 67
Concepts of Ecosystem Ecology A Comparative View
Edited by L.R. Pomeroy and J.J. Alberts
1988. XII, 384p., 93 figures. cloth
ISBN 0-387-96686-2

Volume 68
Stable Isotopes in Ecological Research
Edited by P.W. Rundel, J.R. Ehleringer and K.A. Nagy
1989. XVI, 544p., 164 figures. cloth
ISBN 0-387-96712-5

Volume 69
Vertebrates in Complex Tropical Systems
Edited by M.L. Harmelin-Vivien and F. Bourliere
1989. XI, 200p., 17 figures. cloth
ISBN 0-387-96740-0

Volume 70
The Northern Forest Border in Canada and Alaska
By James A. Larsen
1989. XVI. 272p., 73 figures. cloth
ISBN 0-387-96753-2

Volume 71
Tidal Flat Estuaries: Simulation and Analysis of the Ems Estuary
Edited by J. Baretta and P. Ruardij
1988. XVII, 353p., 1234 figures. cloth
ISBN 3-540-19323-5

Volume 72
Acidic Depostion and Forest Soils
By D. Binkley, C.T. Driscoll, H.L. Allen, P. Schoeneberger, and D. McAvoy
1989. 152p., 34 figures. cloth
ISBN 0-387-96889-X

D. Binkley C.T. Driscoll H.L. Allen
P. Schoeneberger D. McAvoy

Acidic Deposition and Forest Soils

Context and Case Studies of the Southeastern United States

With 34 Figures

Springer-Verlag
New York Berlin Heidelberg
London Paris Tokyo

Dan Binkley
Department of Forest and Wood Sciences
Colorado State University
Fort Collins, Colorado 80523
USA

Charles T. Driscoll
Department of Civil Engineering
Syracuse University
Syracuse, New York 13244
USA

H. Lee Allen
Department of Forestry
North Carolina State University
Raleigh, North Carolina 27695
USA

Philip Schoeneberger
Department of Soil Science
North Carolina State University
Raleigh, North Carolina 27695
USA

Drew McAvoy
Department of Civil Engineering
Syracuse University
Syracuse, New York 13244
USA

Library of Congress Cataloging-in-Publication Data

Acidic deposition and forest soils: Context and case studies of the southeastern United States
Dan Binkley . . . [et al.].
p. cm.—(Ecological studies; v. 72)
Bibliography: p.
Includes index.

1. Acid deposition—Environmental aspects—Southern States.
2. Forest soils—Southern States. 3. Soil acidity—Southern States.
I. Binkley, Dan. II. Series.
QH545.A17I48 1988
574.5′2642—dc19 88-29767

Printed on acid-free paper

Softcover reprint of the hardcover 1st edition 1989

Typeset by Asco Trade Typesetting Ltd., Hong Kong.

9 8 7 6 5 4 3 2 1

ISBN-13: 978-1-4612-8167-2 e-ISBN-13: 978-1-4612-3586-6

DOI: 10.1007/978-1-4612-3586-6

Preface

Knowledge in the field of acidic deposition is expanding rapidly, and both experts and non-experts are challenged to keep up with the latest information. We designed our assessment to include both the basic foundation needed by non-experts and the detailed information needed by experts. Our assessment includes background information on acidic deposition (Chapter 1), an in-depth discussion of the nature of soil acidity and ecosystem H^+ budgets (Chapter 2), and a summary of rates of deposition in the Southeastern U.S. (Chapter 3). A discussion of the nature of forest soils in the region (Chapter 4) is followed by an overview of previous assessments of soil sensitivity to acidification (Chapter 5). The potential impacts of acidic deposition on forest nutrition are described in the context of the degree of current nutrient limitation on forest productivity (Chapter 6). The results of simulations with the MAGIC model provided evaluations of the likely sensitivity of a variety of soils representative of forest soils in the South (Chapter 7), as well as a test of soil sensitivity criteria. Our synthesis and recommendations for research (Chapter 8) also serve as an executive summary.

A complementary volume in the Springer-Verlag Ecological Studies series should be consulted for information on European forests. This volume, Acidic Deposition and Forest Decline in the Fictelgebirge, edited by E.-D. Schultze and O.L. Lange, also provides greater detail on the physiologic responses of trees than we present in our regional assessment.

Our efforts to provide an up-to-date summary were aided greatly by help from: A. Bartuska, W. FcFee, G. Will, E. Cowling, D. Johnson, L. Pitelka, A.

Lucier, K. Knoerr, C. Brandt, R. Harrison, D. Richter, J. Woodman, P. Ryan, J. Waide, W. Swank, J. Turner, D. Valentine, U. Valentine, J. Cosby, J. Pye, G. Hornberger, P. Wigington, Jr., and S. Anderson.

This assessment was supported by funds provided by the Southeastern Forest Experiment Station (Southern Commercial Forest Research Cooperative), within the joint U.S. Environmental Protection Agency—U.S.D.A. Forest Service Forest Response Program. The Forest Response Program is part of the National Acid Precipitation Assessment Program. This book has not been subjected to policy review by the EPA or Forest Service, and it should not be construed to represent the policies of either Agency. Parts of the study were also supported by the Electric Power Research Institute's Integrated Forest Study through Oak Ridge National Laboratory.

Dan Binkley
Charles T. Driscoll
H. Lee Allen
Philip Schoeneberger
Drew McAvoy

Contents

1. Introduction

The precipitation falling in the eastern United States contains higher concentrations of sulfuric and nitric acids than in less polluted areas. Precipitation is normally acidic even in unpolluted areas, where the carbon dioxide in the atmosphere dissolves in water to form carbonic acid, H_2CO_3. Unpolluted precipitation also contains slight quantities of nitric acid (formed by lightning and by microbial activity) and sulfuric acid (formed from naturally occurring sulfur dioxide) as well as alkaline chemicals. Concerns about acidic precipitation focus on the quantity of acidity entering ecosystems rather than the simple presence of sulfuric and nitric acids.

As discussed later, the issues in acid precipitation include a wide variety of concerns, ranging from human health to lake acidification to forest growth. The U.S. government is sponsoring a wide array of research in these areas. The possible impact of acid precipitation on forests falls within the Forest Response Program, which is divided into several regional research cooperatives. The commercial forest lands in the South fall within the Southern Commercial Forest Research Cooperative (SCFRC), which conducts research on many aspects of forest health and growth and the possible impacts of air pollutants. This region, stretching southward from Virginia and Tennessee through Arkansas and eastern Texas, contains a highly diverse array of agricultural and forested ecosystems growing on a wide variety of soils. The specific questions addressed by the SCFRC are:

1. Is there a significant problem of forest damage in the South that might be caused by acidic deposition, alone or in combination with other pollutants?
2. What is the causal relationship between acidic deposition, alone or in combination with other pollutants, and forest damage?
3. What is the dose-response relationship between acidic deposition, alone or in combination with other pollutants, and forest damage?

Purpose of the Assessment

Scientists involved in the SCFRC anticipated that the greatest risk to commercial forests in the South would be from gaseous pollutants (especially ozone) or perhaps from direct effects of acidic deposition on leaves. The soils in the region are generally very old and contain large amounts of clay, which leads to the expectation that acidic deposition poses little threat to the fertility of forest soils. Tabatabai (1985) stated, "It is clear that since forest soils are usually acidic in reaction, measurable effects are not possible." Although the SCFRC scientists expected that possible impacts of acidic deposition would not originate from effects on soils, they felt a critical assessment of the likelihood of soil impacts should be done to gauge the degree of certainty. Therefore, this assessment was initiated to evaluate the potential susceptibility of forest soils in the South to chemical change as a result of sulfur and nitrogen deposition.

Structure of the Assessment

Our assessment is structured to provide both an introduction to readers who are unfamiliar with some of the basic aspects of ecosystem acidity as well as a detailed discussion of possible impacts of acidic deposition on southern forest soils. Experts in the field may want to focus on chapters 4–8, others should find the first three background chapters helpful.

In the rest of this chapter, we describe why the rates of deposition of sulfur and nitrogen are important and how forests may be affected. In chapter 2, we provide an overview of the nature of soil acidity and the dynamics of H^+ in natural nutrient cycling processes and acidic deposition. The current rates of deposition of sulfur and nitrogen in the South are discussed in chapter 3 along with historical trends. In chapter 4 we then characterize the major types of soils found in the South as a foundation for considering how readily they may be affected by acidic deposition. Previous studies have addressed the possible impacts of acidic deposition on forest soils in the eastern United States; in chapter 5 these investigations are reviewed, and we discuss choices of criteria for judging soil sensitivity. The impact of acidic deposition depends strongly on the nutritional status of the forest, and in chapter 6 the state of knowledge on sulfur, nitrogen, and nutrient cations is summarized for commercial forests in the South. To attempt to quantify soil acidification in the South and to evaluate previously proposed sensitivity classes of soils we used an acidification model to simulate changes in soil and soil solution chemistry that may occur for southern soils over the next 140 yr. This analysis is summarized in chapter 7. The final chapter synthesizes the evidence available and describes the level of confidence we have in our expectations of the potential impacts of acidic deposition on forest soils in the South.

Background on Why This Has Become an Important Topic

The historical development of scientific understanding and public concern over acidic deposition was reviewed by Cowling (1982), and much of our summary comes from his work. The harmful effects of severe pollution were first recognized in Europe at least three centuries ago, but modern scientific investigations began about one century ago with the work of Robert Smith in England. In 1852, Smith described the chemistry of rain in the vicinity of Manchester, England. In remote fields, he found that ammonium carbonate (probably ammonium bicarbonate, NH_4HCO_3) dominated rain chemistry. This most likely formed when ammonia from animal wastes combined with natural carbonic acid in the air, neutralizing the acidity to form an ammonium bicarbonate salt. Nearer the city, a mixture of ammonia from animal wastes and sulfuric acids from coal burning combined to produce the ammonium sulfate salt [$(NH_4)_2SO_4$]. Within the city, apparently the production of sulfuric acid exceeded the ammonia available to neutralize it, and sulfuric acid dominated rain chemistry. In 1872, Robert Smith (cited in Cowling 1982) published a classic book, "Air and Rain: The Beginnings of a Chemical Climatology," that included discussions of effects of coal combustion, wind trajectories, effects of sea winds, and precipitation patterns. However, Smith's book had little influence in the decades that followed, and the acidic nature of precipitation received little attention until the 1950s.

Eville Gorham (1955) described the chemistry of rain and concluded that acids in rain were derived from industrial emissions and that lakes, streams, and soils could be acidified by these acids. In Europe, Hans Egner was interested in the fertilization effect of chemicals in rainfall, and he initiated a large-scale network of collecting stations to characterize the chemistry of rain in Sweden. This network expanded to other countries in Europe, and most of that region now has a record of rainfall chemistry for 30 yr or more. Some regional measures of rain chemistry across the United States were made in the 1950s (such as Junge and Werby 1958), but a permanent network was not begun until the middle-to-late 1970s.

In the early 1960s in Sweden, Svante Oden compared data from the precipitation chemistry network with regional information on the chemistry of streams and lakes. He concluded that acidic deposition was leading to acidification of aquatic ecosystems, and he stressed that owing to the transport of air pollutants across international boundaries it was a global problem. By 1972, concern among Swedish researchers led to a presentation at the UN Conference on the Human Environment titled "Air Pollution Across National Boundaries: The Impact on the Environment of Sulfur in Air and Precipitation" (Bolin et al. 1972). In the same year, the Norwegian government initiated the SNSF project (the Norwegian Interdisciplinary Research Programme titled "Acidic Deposition: Effects on Forest and Fish"), with an annual budget of about $2 million/yr for 8 yr.

Concern in the United States developed in the 1970s, spurred by lectures from Swedish scientists and by the findings of U.S. researchers on the strongly acidic nature of precipitation in the northeastern United States. In 1975, the first international conference on acidic deposition was held at Ohio State University (Dochinger and Seliga 1976). In 1979, U.S. President Carter initiated a 10-yr research program to determine the causes and impacts of acidic deposition in the United States.

Lakes Versus Forests

Current concern (and research) focuses on possible effects of acid deposition on both aquatic and terrestrial ecosystems, but these two types of ecosystems differ in several ways. The pH of lakes is often in the slightly acid range of 5.5 to 6.5; forest soils are commonly strongly acidic with pH levels of 4 to 6. Despite the higher pH values, lakes are usually much more sensitive to changes in chemistry than are soils. Lakes are often dilute solutions, and small changes in inputs or release/retention of solutes can readily change their overall chemistry. The chemistry of soil water is strongly affected by the chemistry of the soil; given the very large pools of chemicals with soils, changes in solution inputs and outputs are strongly buffered. The chemistry of water in lakes is highly dynamic because of short-term changes in the amount and chemistry of water running off from the surrounding land; the average residence time of water in lakes may range from only a few days to several years. In constrast, the average residence time of water in soils is even shorter (days to weeks), but strong interactions with the soil matrix tend to buffer the chemistry of drainage waters.

Concerns over Effects on Forests

Historic Understanding of pH and Tree Growth

The linkage between soil pH and plant growth and health have been appreciated for at least 2000 yr. Tisdale et al. (1985) noted that liming of soils to improve plant growth was recommended during the Golden Age of Greece and during the Roman Empire. As early as 1836, F. Unger grouped vegetation into classes based on their tolerance of soil acidity. Lime-loving plants were found in high-pH soils, whereas lime-avoiding plants thrived in acidic, low-pH soils. About 100 yr later, hydroponic studies indicated that plants were not very sensitive to the pH of solution in the root environment; Arnon and Johnson (1942) showed that most plants grew equally well in solutions ranging in pH from 3 to 9. The relationships between soil pH and tree growth result from the indirect effects of soil acidity on other factors that influence trees. These factors include:

1. *Aluminum concentration in soil solution*. Acidic soils often contain elevated concentrations of aluminum associated with the exchange complex (some organic soils with low pH may be exceptions). Such soils, in turn, may have high concentrations of aluminum in the soil solution. The root activity of some plants is impaired by high concentrations of aluminum. Many tree species commonly grow on very acidic soils and appear tolerant of high aluminum concentrations in soil solutions. Such species may still be sensitive if the ratio of aluminum to nutrient cations is high in soil solutions. This condition may limit root uptake of calcium and magnesium, thus creating nutrient deficiencies. A great deal of research is currently focused on these concerns.
2. *Solubility of nutrients*. Some important plant nutrients, such as phosphate and molybdenum, are anions that exhibit reduced solubility or increased adsorption at low pH. Reductions in solution pH owing to acidification may decrease the availability of these nutrients for forest growth. Unfortunately, processes regulating the aqueous concentrations of phosphate and molybdenum are not well established (see chap. 6).
3. *Indirect effects via impacts on microbes*. Trees depend on microbes for a host of functions, from decomposer microbes that recycle nutrients to mycorrhizal fungi that aid in nutrient uptake by plants. The microbial community is dynamic, thus changes in soil pH can alter both the relative dominance of microbe types and the rates of microbial processes. Myrold (1987) reviewed the probable impacts of acid deposition on soil biota and concluded that deposition rates of < 1 kmol H^+/ha annually (see table 1.1 for a summary of measurement units commonly used in acidic deposition discussions) are unlikely to have much effect but that effects may begin to appear at higher rates. Perhaps more critical than the soil biota as a whole are mycorrhizal associations with tree roots. The review of Cline et al. (1987) determined that there is no evidence to support ideas of impaired mycorrhizal associations at current deposition rates in the eastern United States. Because these reviews concluded that effects on soil biota are not likely to be of primary importance (but see also W.H. Smith's [1987] review of impacts on the rhizophere), our assessment focuses on soil chemistry and tree nutrition.

Forest Decline in Europe

Public concern over acidic deposition became focused in the early 1980s (see Neuman 1986) with reports from Europe of widespread decline of forests. The decline and death of the forests did not appear related to easily diagnosed diseases or nutritional problems, and the word *Waldsterben* (forest death) was used to describe the novel symptomology of forest decline (Schutt et al. 1983, Schutt and Cowling 1985). A survey of forests in West Germany (Federal Republic of Germany [FRG]) in 1984 concluded that about half the total forest area showed symptoms of *Waldsterben* (Schutt and Cowling 1985). Since then, many low-elevation stands of Norway spruce (*Picea abies* L.) have recovered, and the decline of some other stands has been shown to be simple nutrient deficiency rather than *Waldsterben* (C. Brandt, D. Johnson, L. Pitelka, pers. comm.). Some German mensurationists conclude there is no evidence of widespread forest decline

Table 1.1. Measurement units

Unit	Definition
mol, mole	The basic unit of chemistry, equal to the number of atoms of hydrogen present in 1 g (6.02×10^{23}).
molecular weight	The weight of a mole of molecules of any chemical. Sodium weighs about 23 g/mol, calcium about 40 g/mol.
mol_c	A mole of charge, either positive (cations) or negative (anions). A mol_c of H^+ is the same as a mol of H^+ because it contains only 1 charge/ion. A mol_c of Ca^{2+} (20 g/mol_c) equals only $\frac{1}{2}$ mol of Ca^{2+} because each ion has two charges.
mol/L	A mole of a chemical in a liter (usually in water). A solution containing 1 mol/L of nitric acid has 1 g of H^+ and 62 g of NO_3^- (which is 48 g of oxygen and 14 g of nitrogen).
mol_c/L	A mole of ion charge per liter. A solution containing 1 mol of H_2SO_4 has 2 mol_c/L of H^+ and SO_4^{2-}.
mmol, $mmol_c$	One-thousandth (10^{-3}) of a mol or of a mol of charge.
μmol, μmol_c	One-millionth (10^{-6}) of a mol or of a mol of charge.
g SO_4^{2-}	Grams of sulfate, including the weight of oxygen.
g SO_4^{2-}—S	Grams of sulfur only, present in the form of sulfate.
kg NO_3^-	Kilograms (10^3 g) of nitrate, including the weight of oxygen.
Mg NO_3^-—N	Megagrams (10^6 g, 10^3 kg) of nitrogen only, in the form of nitrate.
ha	Hectare (10^4 m^2), equal to 2.47 acres.
kmol/ha, $kmol_c$/ha	A kilomole (10^3 mol) or kilomole of charge per hectare. 1000 mm of rain (39 in.) at pH 5.0 contains 0.1 $kmol_c$/ha of H^+; the same amount at pH 4.0 contains 1.0 $kmol_c$/ha.

(G. Will, pers. comm., 1987). However, the 1987 European Economic Community (EEC) survey found that from 2% (Sweden) to 29% (FRG) of the Western European conifer forests showed 25% or greater loss of foliage (C. Brandt, pers. comm.). Ellis Cowling (1987, pers. comm.) summarized the current ideas in Germany on causes of forest decline:

Natural stresses are still discussed as contributing factors to novel forest decline, perhaps as part of ordinary cycles of pests and pathogens. However, the evidence is not clearly supportive.

Viruses and virus-like organisms have been isolated from damaged trees, but they do not appear pathogenic and this idea is no longer considered a major contributor to forest decline.

Ozone alone has not been found to produce the symptoms observed in declining forests, even when experiments combine ozone exposure with acid treatments. The possible role of ozone is therefore unclear.

Photochemical oxidants such as hydrogen peroxide continue to be viable candidates in causing forest decline, but more conclusive work is needed.

Magnesium deficiency has been observed in Norway spruce stands, and fertilization with magnesium alleviates symptoms of decline. No connection has been established between Mg deficiency and acidic deposition, and Mg deficiency does not account for the decline of other species (or of Norway spruce on all sites).
Acidification/aluminum toxicity may explain some declining stands in the Solling area near Göttingen, but direct connections to acidic deposition have not been demonstrated, and this mechanism does not explain decline in other areas.
Excess nitrogen is under serious consideration as a mechanism leading to problems with frost hardiness and the nutritional balance of trees, but strong evidence is not available.
Organic chemicals in precipitation are being examined as possible causes of growth alteration in trees, but the research has focused only on identification of chemicals in the precipitation and growth effects have not been examined.
General stress hypothesis anticipates that decline symptoms arise from the interactions of several stresses. This hypothesis is appealing, but experimentally intractable; it is discussed more than it is tested.

Cowling also notes that little disruption in the timber markets has occured in Europe; about 3% of the annual harvest in Germany comes from the salvage logging of *Waldsterben* sites. The current situation may, of course, change in the future.

Forest Decline in the United States

No widespread declines of forests in the United States are documented, but a variety of forest health problems have aroused concern. In New England, red spruce (*P. rubens* Sarg.) on the top of Camels Hump in Vermont exhibited marked mortality in the late 1970s, and the basal area of the forest continued to decline into the 1980s. Two suggested major causes (see Johnson and McLaughlin 1986) are:

Deposition of acids and heavy metals, proposed mainly because no obvious natural causes are evident and because large-scale forest declines have been observed only at high elevations where deposition rates are much greater than in other areas. However, there is no strong evidence that acidic deposition is contributing to the decline of high-elevation red spruce stands, and previous suggestions of excessive nitrogen contents of needles leading to poor frost hardiness have not been supported (Friedland et al. 1985, 1988).
Climate fluctuations, which appear to be an important contributing factor. Severe winters have been associated with multiple periods of decline in red spruce (A. Johnson and McLaughlin 1986, A. Johnson and Friedland 1986).

In parts of the Southern Appalachians, both red spruce and Fraser fir [*Abies fraseri* (Pursh) Poir.] show signs of impaired health. Until recently, the widespread death of fir trees was attributed to an exotic insect, the balsam woolly adelgid (*Adelges piceae* Ratz.). However, Bruck and Robarge (1987) noticed in 1983 that many spruce trees mixed with the dying fir were also dead or dying. These researchers cored spruce trees at high and low elevations and found that high-elevation trees began exhibiting growth declines in the 1960s. Similar findings were also reported by Winner et al. (1986) for the Great Smoky Mountains growth. The adelgid does not attack spruce trees, so the decline of the spruce must relate either to the microenvironmental effects of the deaths of neighboring firs or to some direct mechanism. Current research has found very high rates of nitrogen availability in soils of spruce and fir stands in the Southern Appalachians (Strader et al. 1988) as well as high rates of nitrate leaching (> 0.8 kmol-N/ha annually) from soils (D.W. Johnson et al. 1988). However, connections with forest decline are speculative at this point, and some researchers stress the importance of natural patterns of stand development (see Zedaker et al. 1986, Witter and Ragenovich 1986; but see also McLaughlin and Adams 1987).

A widespread reduction in the growth rates of natural pine forests in parts of the South from the 1940s through the 1980s was reported by Sheffield et al. (1985). Plantations, especially on industrial lands, showed no decline. Possible causes for the decline were identified as: atmospheric deposition, increased stand density, increased stand age, increased competition from hardwoods, drought, lowered water tables, loss of old-field conditions, and increased incidence of diseases. No evidence implicating acidic deposition was available. The absence of growth declines on intensively managed plantations in the Coastal Plain indicated either that atmospheric deposition was not a factor or that management regimes countered any effects of deposition.

Pathways of Acidic Deposition Impacts

Direct Acid Damage to Leaves

It seems reasonable that direct deposition of strong acids onto leaf surfaces may result in damage. A wide variety of studies using many approaches (generally some application of varying pH solutions to the surfaces of leaves of seedlings) has tested this hypothesis. Three mechanisms seem to be most significant: leaf cutical erosion, leaf cell acidification, and foliar leaching.

Leaf Cutical Erosion

Erosion of the leaf cuticle has been documented with scanning electron microscopy (see Shriner 1977). This effect occurs only when leaves are treated with very acidic solutions ($pH < 3.5$). Erosion of the cuticle could also make leaves more susceptible to damage from gaseous pollutants.

Leaf Cell Acidification

Acidification of the leaf cells was implicated in a study by Paparozzi and Tukey (1984). They found that very acidic treatments ($pH \leqslant 3.2$) caused collapse of palisade and epidermal cells without eroding the cuticle. They speculated the direct cause could have been acid hydrolysis of cell walls, perhaps coupled with the osmotic stress of the high-ionic-strength solution outside the cells.

Foliar Leaching

Leaching of nutrients and organic chemicals is usually cited as the most likely cause of harm to intact trees receiving acidic deposition at current, ambient rates. Throughfall (precipitation dripping through canopies) in many forests is commonly higher in pH and enriched in nutrient cations relative to incident precipitation (G.G. Parker 1983, 1987), suggesting the removal of nutrient cations from the leaf surfaces. Many studies have also treated trees (or even irrigated entire canopies) by mist application of acidic solutions, and then measured the ionic composition of the throughfall. Most of these studies have also documented increased losses of nutrient cations from foliage, but the concentrations of these cations in the leaves usually remain unchanged. Researchers have often assumed that inputs of H^+ deplete these cations from ion exchange sites on or in the leaves. However, Cronan (1984) demonstrated that during the growing season about half of the neutralization of acidity in throughfall was accomplished by production of bicarbonate or low-molecular-weight organic acids.

Nutrient cations rarely limit the growth of forests, but the limitations can be substantial where they do occur. In Europe, some Norway spruce stands have been shown to have low concentrations of Mg in foliage and growth increases after fertilization with Mg (Zöttl and Hüttl 1986). If acidic deposition leached Mg from needles on Mg-deficient sites, it could exacerbate a preexisting stress. Some red pine stands on glacial outwash sands in New York are highly deficient in potassium (Comerford et al. 1980); the degree to which this problem is enhanced by strong acid deposition has not been examined. See chapter 6 for a discussion of cation limitations in the South.

Fertilization Through Foliar Uptake or Soil

Sulfur and nitrogen are essential plant nutrients that may limit forest growth, so additions of these elements may improve tree nutrition. Indeed, many experimental additions of sulfuric and nitric acids to tree seedlings have stimulated growth rather than inhibited it (for a review, see Morrison 1984). It might appear that the net effect of S and N deposition is balanced between harmful effects of the acidity and beneficial effects of the sulfate and nitrate ions (see Morrison 1984). However, if plants assimilate the sulfate or nitrate, they neutralize the associated acidity (as described in chap. 2) and prevent any acidification. However, problems may still occur when acids directly damage leaves prior to assimilation of the nutrient anions.

Many plants have the ability to assimilate nutrient ions through foliage. Throughfall often contains less nitrate than does precipitation, indicating net uptake by the canopy (Parker 1983). Forests can even be fertilized by applications of liquid fertilizers to canopies (see R. Miller 1979), but this method is generally cost effective only for orchards. However, not all tree species are capable of assimilating nitrate through foliage because they lack the necessary enzyme (nitrate reductase) (Smirnoff et al. 1984). For example, Bowden et al. (1987) concluded that high-elevation spruce forests probably assimilate only about 7 mmol-N/m^2 (0.1 g-N/m^2) annually through foliage.

Throughfall is usually enriched in sulfate relative to precipitation (G.G. Parker 1983, 1987), reflecting wash-off of sulfate deposited onto the canopy. Most plants can assimilate sulfate through leaves, and direct uptake of sulfur dioxide also occurs (Mengel and Kirkby 1982). Sulfur limitations of forest growth are rare, however, and the nutritional value of deposited sulfur is probably negligible (see chap. 6).

The deposition of nitric acid onto forest soils undoubtedly benefits forests with limited supplies of nitrogen. Very high rates of nitrogen deposition, such as occur in parts of Europe, have probably removed any nitrogen limitation on growth. Across the southern United States, most forests are strongly limited in growth by nitrogen availability, so deposition of nitric acid provides a clear opportunity for increased forest growth.

Forests require only about 3% (on an atom basis) as much sulfur as nitrogen, and about twice as much sulfur is deposited in the South as nitrogen. Therefore, sulfur limitations are rare and sulfur deposition probably provides no nutritional benefits to southern forests (see chap. 6).

Acidification of Soil

If the sulfate and nitrate in acidic deposition are not assimilated by plants and microbes, the input of H^+ has the potential to increase soil acidity and lower soil pH. Subsequent effects could include:

—Altered microbial populations and activities.
—Reduction in the solubility of phosphorus and molybdenum.
—Increased solubility of potentially toxic metals, such as aluminum and manganese.

If the sulfate and nitrate are transported from the soil with drainage water, the leaching of cations would be enhanced. Loss of these essential cations may lead to the potential decline of forest growth. Where exchange complexes are dominated by aluminum, export of aluminum can also pose a threat to aquatic ecosystems.

For these reasons, assessments of the impacts of acidic deposition need to consider the quantity of materials deposited, the fate of the sulfate and nitrate, the properties of soils within watersheds, and downstream aquatic ecosystems.

The Uncertain Nature of Truth in Ecosystem Studies

Woodman (1987) reviewed the "facts and suspicions" of pollution-induced injury in North American forests, and concluded that, "there is no rigorous proof that regional pollution has changed the average growth rates or development of any North American forest." Woodman maintained that causal connections between pollution and tree responses require demonstration that:

—Tree injury or dysfunction must be associated with the presence of the suspected pollutant(s).

—Tree injury or dysfunction must be produced when healthy trees are exposed to the suspected pollutant(s) under controlled conditions.
—Variation in the degree of injury to trees in the forest must be duplicated when clones of the same trees are exposed to the suspected pollutant(s) under controlled conditions.

The first two postulates are necessary to draw a firm inference, whereas the third is further evidence of cause and effect. Given the widespread nature of air pollution in regions of North America, the first of Woodman's conditions is easily met. The second and third conditions may be met when the stress produced by exposure to a pollutant under controlled conditions is severe, but controlled experiments cannot incorporate the diversity of natural stresses found in real ecosystems. These perspectives need to be evaluated in relation to the nature of the predictability of ecosystem functions and patterns as well as the nature of decision making under uncertainty.

Bormann (1985) nicely summarized the nature of stress in forests and the likely response of forests to increased stress. He emphasized that most (if not all) forests experience continual stress from environmental conditions and competition and that any effect of air pollutants needs to be considered as additional stress rather than the initiation of stress. Bormann emphasized that:

—Our knowledge of actual pollution inputs is fragmentary.
—Long-term ecosystem data are rare.
—Even if growth or population trends are detected, they are difficult to relate in a quantitative way to specific air pollutants.

Bormann noted that even in areas of clear forest decline, such as spruce forests in Europe, and high-elevation red spruce forests in the United States, causal factors have not been clearly identified despite very intensive research efforts. Given this, he asks how reasonable it would be to expct clear causal proof of more subtle effects (such as a 10% growth reduction) that might result from pollution.

Classical hypothesis testing in science emphasizes the importance of strong evidence that a hypothesis is true before accepting it, even at the risk of accepting a null hypothesis that may be untrue. This approach is echoed in the conclusions of Woodman (1987). However, important changes can occur in forest without opportunities for such rigorous hypothesis testing; classical proof from hypothesis tests may be obtainable only in the most severe cases (but even this has yet to be demonstrated in current cases of forest decline). Indeed, it is unclear which would be the more reasonable null hypothesis. Is it safer to assume that pollutants have no effect until proven otherwise or that pollutants that are known to be potentially harmful are likely to be harmful until proven otherwise?

Policy decisions are often based on imperfect data (Grobstein 1983), simply because certainty is an unattainable goal. In our assessment, we summarize the current levels of uncertainty about the susceptibility of forest soils in the South to acidic deposition and discuss the likely reduction in uncertainty that could be obtained from further research.

2. The Nature of Soil Acidity and H^+ Budgets

Many processes affect the acidity of soils and soil solutions. It turn, the acidity of soil solutions affects many processes. In this chapter, we review the nature of soil acidity, and synthesize the major processes involved in the net generation and consumption of acidity in forest. Two case studies of loblolly pine (*Pinus taeda* L.) forests provide examples of how H^+ budgets are linked with rates of soil acidification.

Because the concentration of H^+ in soil solutions commonly varies by orders of magnitude, its concentration is commonly expressed on a logarithmic pH scale; pH is defined as the negative of the logarithm of activity (~concentration) of H^+. A solution with a pH of 7 has a H^+ of 10^{-7} mol/L. At this pH, the concentration of OH^- ions is also 10^{-7} mol/L, and the solution is considered to have a neutral pH. A pH above 7 means the concentration of H^+ is less than 10^{-7} mol/L, and the solution is considered alkaline. Acidic conditions prevail where the pH is less than 7 and the concentration of H^+ exceeds 10^{-7} mol/L.

A second aspect of acidity is the ability of a solution to resist changes in pH when H^+ is added or removed. The pH of a solution describes the *intensity* factor of acidity, and the ability to resist changes is a *capacity* factor. Therefore, a complete characterization of the acidity of a soil or solution involves the current pH, and how much the pH changes when H^+ is added and when H^+ is removed (= OH^- added). The resistance to pH change as H^+ is added is called acid neutralizing capacity (ANC), or titratable alkalinity. The resistance to pH change as OH^- is added is base neutralizing capacity (BNC), or titratable acidity.

The pH of precipitation in unpolluted regions is commonly considered to be about 5.6 owing to the dissolution of carbon dioxide in the air ($10^{-3.5}$ atm, 320 mg/kg) with water to form carbonic acid (the true pH of unpolluted rain is usually somewhat less than 5.6). A pH of 5.6 means the concentration of H^+ is about $10^{-5.6}$ mol/L. Precipitation with a pH of 4.6 would have ten times more H^+ than at pH 5.6. In polluted areas, precipitation pH commonly falls in the range of pH 4.2 to 5.0. Forest soils commonly fall within the range of pH 4 to 6, but exceptions to this range are not unusual. Lakes and streams typically have fairly high pH, in the range of 5.5 to 6.5; acidified lakes may have pH values as low as 4.5.

Over the short term, the concentration of H^+ affects many of the processes that regulate ion concentrations in soil solutions. Over longer time scales, the concentration of H^+ depends on total quantities of ions entering and leaving the soil. For example, a forest receiving 1000 mm of precipitation at a pH of 5.6 would receive 25 mol of H^+/ha annually, whereas a forest receiving 2000 mm of precipitation at the same pH would receive twice as many H^+ (50 mol/ha). Although each forest received rain of the same pH, the one receiving more precipitation might acidify more quickly because it received twice as much H^+.

Precipitation and drainage waters also contain carbonic acid that is not dissociated, and this acid provides BNC to the solution. Carbon dioxide dissolves in water to form carbonic acid, some of which dissociates to form H^+ and bicarbonate (HCO_3^-):

$$CO_2 + H_2O \rightarrow H_2CO_3 \rightarrow H^+ + HCO_3^- \quad (1)$$

The proportion of the carbonic acid that dissociates is governed by LeChatlier's principle that states when a system is stressed in one way, it shifts the other way to relieve the stress. In this case, adding H_2CO_3 would "stress" the system, which would relieve the stress by forming more H^+ and HCO_3^-. The tendency of the carbonic acid to dissociate (release H^+) reaches a balance with bicarbonate's tendency to associate (accept H^+). This feature is described by equilibrium constants, symbolized K_{eq}, that relate the tendency of compounds (such as acids) to dissociate to the concentration of the products (dissociated form) divided by the concentration of the reactants (undissociated form). For carbonic acid (at 25° C), the relationship is:

$$K_{eq} = 4.67 \times 10^{-7} = \frac{(H^+)\,(HCO_3^-)}{(H_2CO_3)} \quad (2)$$

The ( ) denote "activities" of the products and reactants, which approach the concentrations of the chemicals in very dilute solutions. At higher concentrations, activities are lower than concentrations. The extent of chemical equilibrium is influenced by temperature and pressure as well as ionic strength; for a detailed discussion, see Stumm and Morgan (1981). For our purposes, we shall assume activities and concentrations are equal and use [ ] to denote chemical concentrations in mol/L.

This simple equilibrium relationship allows calculation of the dynamics of acid/base systems. For example, if the pH of precipitation were 5.6 because of carbonic acid, the concentration of H^+ would be $10^{-5.6}$ mol/L, which also equals 2.5×10^{-6} mol/L. The concentration of bicarbonate would also be $10^{-5.6}$ mol/L, because there would be one bicarbonate formed for every H^+. Knowing these two concentrations allows the equation to be solved for the concentration of undissociated carbonic acid:

$$4.67 \times 10^{-7} = \frac{[10^{-5.6}]\,[10^{-5.6}]}{[H_2CO_3]} \quad (3)$$

The concentration of undissociated carbonic acid must be 1.35×10^{-5} mol/L, and the proportion of the acid that is dissociated is about 16%.

Because carbonic acid does not fully dissociate over the pH range common in water solutions (pH 2 to 12), it is called a "weak" acid. In contrast, nitric acid is considered a strong acid because its very low pK_a (-1) causes essentially complete dissociation even at very low pH values.

In the previous example, doubling the amount of precipitation (at the same pH) doubled the input of H^+. Consider now 1000 mm of pH 5.6 rain falling on two different soils. A soil with pH 5.6 receives 25 mol H^+/ha. However, if the pH of the soil were 6.6, the soil would receive not only the H^+ from the dissociated carbonic acid in the rain, but also the additional H^+ resulting from deprotonation of the acid in the elevated pH of the soil environment. In this case, the higher soil pH would "stress" the solution by having fewer H^+ than the equilibrium solution. The stress would be relieved by further dissociation of the carbonic acid. The extra dissociation can be calculated with the Henderson–Hasselbalch equation, relating the degree of dissociation of an acid to the solution pH and the strength of the acid (in this case expressed as pK, the negative log of K_{eq}):

$$6.60 = 6.33 + \log\frac{[HCO_3^-]}{[H_2CO_3]} \quad (4)$$

At pH 6.6, carbonic acid (with a pK of 6.33) would be 65% dissociated. If 16% dissociation gave an input of 25 mol/ha in 1000 mm of rain, 65% dissociation would provide an input (from the same rain falling on soil with a pH of 6.6) of 100 mol/ha, or four times more.

If the soil pH were 4.6 rather than 5.6 or 6.6, the same type of calculations would show that the dissociation of carbonic acid would be diminished owing to elevated H^+ in the soil solution. In fact, only about 1.6% of the carbonic acid would be dissociated and the quantitative input of H^+ would be only 2.5 mol/ha.

The concentrations of carbon dioxide in the soil are typically much higher than in the atmosphere, and the acid/base chemistry of carbonic acid may be more important in soils than in precipitation (see later discussion).

Thermodynamic calculations are simple and provide important information on the acidification of soils and drainage waters. The important points are:

1. The concentration of H^+ (expressed as pH) is an important master variable that regulates aqueous and solid-phase reactions.
2. The quantity of H^+ added to an ecosystem is also important in the regulation of the long-term acid/base status of soils and drainage water.
3. Weak acid added to a soil may either dissociate more completely or less completely, so the pH of precipitation may not indicate the true amount of H^+ entering an ecosystem in precipitation.

Weak Acid Behavior of Soils

A wide range of parameters are used by soil scientists to characterize the acid/base properties of soil, including the cation exchange capacity (CEC), the effective CEC, the permanent charge CEC, variable charge CEC, the percentage base saturation, exchangeable acidity, nonexchangeable acidity, titratable acidity, and so on. This array of terms can be simplified by considering the soils as mixtures of weak acids that vary in concentration, strengths (pK_a's), and degree of dissociation. For example, permanent charge CEC derives from irregularities in the crystal structure of clays, such that a net negative charge exists on the micelle. The micelle may adsorb H^+, becoming a clay acid (called an argillic acid), or may adsorb a cation such as K^+ and be a dissociated clay (called an argillate). The tendency of H^+ and other cations such as K^+ to occupy the exchange sites follows LeChatlier's principle; if the concentration of a single ion is increased in the soil solution, that ion will tend to associate with the exchange complex. If the concentration of an ion in solution decreases, that ion will tend to be displaced to the solution from the exchange site.

Levine and Ciolkosz (1988) used a very simple representation of competitive ion exchange in their computer model of soil acidification: the ion that occupied the greatest number of exchange sites was the simplest to displace. If the concentrations of all exchangeable cations were equal, then the replacement series was $Na > K > Mg > Ca > Al(H)$.

Most models of soil exchange and acidification use a slightly more complex approach based on chemical equilibrium equations. An equation (such as eq. 5) is written, and an equilibrium expression is developed (eq. 6). Because of the complex nature of systems with dissolved, adsorbed, and solid phases, the "constants" are not true thermodynamic constants; they are commonly referred to as *selectivity coefficients*. Various approaches to calculation of these equilibrium constants are used (Ulrich et al. 1971, Reuss 1983, Cosby et al. 1985a, Goldstein et al. 1985). One approach to calculating cation selectivity is the Kerr equation described here (after Bohn et al. 1985, and Reuss and Johnson 1986). Consider the replacement of Ca^{2+} on the exchange complex (CaX) by aluminum:

$$CaX + \frac{2}{3}Al^{3+} \rightarrow \frac{2}{3}AlX + Ca^{2+} \qquad (5)$$

The equilibrium expression of this reaction is:

$$K_s = \frac{[AlX]^{2/3}\,[Ca^+]}{[CaX]\,[Al^{3+}]^{2/3}} \tag{6}$$

This equation can be rearranged to express the ratio of the ions in the solution as a function of the constant times the ratio of the ions on the exchange complex:

$$\frac{[AlX]^2}{[CaX]^3} = K_s\,\frac{[Al^{3+}]^2}{[Ca^{2+}]^3} \tag{7}$$

(Both sides of the equation were raised to the third power to eliminate the fractional exponents.)

As a short-term approximation, the ratio of the ions on the exchange complex can be considered a constant. Therefore, the concentration of aluminum in the soil solution is approximated by:

$$[Al^{3+}] = K_t[Ca^{2+}]^{3/2} \tag{8}$$

where K_t is the square root of

$$\frac{[AlX]^2}{K_s[CaX]^3} \tag{9}$$

A variety of approaches has been developed for calculating selectivity coefficients (Sposito 1981, Bohn et al. 1985). The Gapon equation uses moles of charge rather than moles; the Gaines–Thomas equation uses equivalent fractions on the exchange complex (and equivalent fractions have no thermodynamic validity [Sposito 1981]); and the Vanselow equation incorporates a more realistic evaluation of adsorbed ions as "homogenous mixtures" of solids rather than as dissolved species.

The concept of selectivity coefficients is simple, and in general terms it is useful. For example, equation 9 shows that if the concentration of all cations in the soil solution increased, aluminum would increase more (to the 3/2 power) than would calcium. This pattern explains why increasing the salt concentration (ionic strength) of soil solutions tends to lower pH and increase the solution concentration of aluminum more than divalent and monovalent cations. If the selectivity coefficient is known, it is also possible to calculate the change in solution aluminum concentrations as the exchangeable calcium (or base saturation) changes. However, the anionic charges on exchange complexes do not behave as free ions in solution, and there are conceptual and empirical problems in correctly accounting for the activity of solid-phase ions.

The disequilibrium of soil solutions coupled with our poor understanding of soil-solute interactions of many ions makes it very difficult to quantify these dynamics precisely (see Bohn et al. 1985, Reuss and Johnson 1986). In reality, selectivity coefficients (K_s) are not true equilibrium constants, and they vary with the ionic strength of soil solutions and the distributions of ions in solution and on

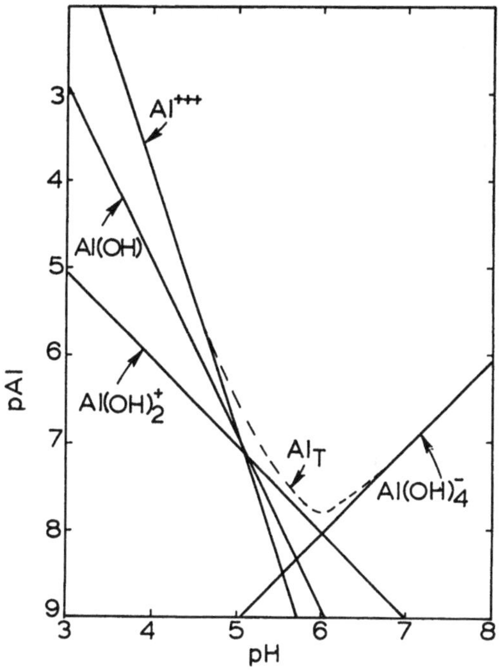

Figure 2.1. Aluminum concentration [pAl = −log (Al)] and speciation as a function of pH.

the exchange complex. However, the application of the simple chemical equilibrium concepts to soil chemistry may provide a useful approximation of expected trends.

Another complication of the behavior of soils as weak acids arises from the fact that H^+ does not actually adsorb to exchange sites on mineral clay surfaces. Adsorbed H^+ is unstable and partially dissolves the clay, releasing aluminum. However, aluminum is considered to be an acidic cation because it behaves as an acid by donating H^+. This acidic character may be attributed to the high charge-to-size ratio of aluminum causing six water molecules to bind or complex to each aluminum ion. Both the total solubility and the degree of protonation of aluminum species (fig. 2.1) depend upon pH:

$$Al^{3+} + 6H_2O \rightarrow [Al(H_2O)_6]^{3+} + H^+ \quad (pH<4.5) \qquad (10)$$
$$\longleftrightarrow [Al(H_2O)_5(OH)]^{2+} + H^+ \quad (pH\ 4.5\text{–}5.0) \qquad (11)$$
$$\longleftrightarrow [Al(H_2O)_4(OH)_2]^{+} + H^+ \quad (pH\ 5.0\text{–}6.5) \qquad (12)$$
$$\longleftrightarrow [Al(H_2O)_3(OH)_3] + H^+ \quad (pH\ 6.5\text{–}8.0) \qquad (13)$$
$$\longleftrightarrow [Al(H_2O)_2(OH)_4]^{-} + H^+ \quad (pH>8.0) \qquad (14)$$

As pH increases, aluminum hydrolyzes, releasing H^+ to the soil solution, behaving as an acid. However, as pH decreases, aluminum consumes H^+ and acts as a base. Thus aluminum may act as a weak acid or weak base, just like all other weak acids and bases.

A soil with its permanent charge complex dominated by aluminum will have the pH of the soil solution governed largely by the acid/base chemistry of aluminum and will tend to have a low pH. If the exchange complex is dominated by basic cations such as Ca^{2+} and K^+, the exchange complex will behave as a dissociated acid, and exhibit a higher pH. This pool of aluminum is referred to as *exchangeable-titratable acidity*, which can be analytically determined by displacing aluminum into solution by extraction of the soil with potassium chloride, and then titration of the extract by a strong base (Bohn et al. 1985).

The next component of the soil exchange complex is the variable charge pool, which consists mostly of organic acid groups contained in soil organic matter. Under common pH values in forest soils, the carboxyl groups (R-COOH) comprise most of this pool. At low pH, most of the carboxyl groups remain undissociated; as pH increases, carboxyl groups donate H^+ to the soil solution. The negatively charged carboxylates may then adsorb other cations, such as Al^{3+}, Ca^{2+}, and K^+. This represents the *nonexchangeable-titratable pool* of soil acidity (and variable charge CEC) because it is not extracted with potassium chloride solutions but does supply H^+ when the soil is titrated with a strong base. Some aluminum may also not be extractable by potassium chloride, and therefore it comprises part of this pool.

A final pool of potential acidity in forest soil are the vast stores of reduced C, N, and S. These pools represent Lewis alkalinity and when oxidized by oxygen (a Lewis acid), Bronsted/Lowry acids are produced (carbonic acid, nitric acid, and sulfuric acid). The rate of C oxidation in forests is often 500 kmol/ha annually or greater, yielding 500 kmol/ha or more of carbonic acid. Fortunately, most of the carbonic acid degases into the atmosphere. Rates of N oxidation in forests commonly range from 0 to 10 kmol/ha annually, producing from 0 to 10 kmol/ha of nitric acid. Much of the nitrate is recycled into microbes and vegetation, where N reduction consumes the acidity. If N uptake is less than N oxidation, soil acidification may result. The rate of S oxidation is much lower than N oxidation in forests, but it may still be important in some cases.

Soil scientists use the term *base saturation* to describe the fraction of the soil exchange complex that is occupied by cations such as Ca^{2+}, Mg^{2+}, K^+, and Na^+. These cations are not bases in a chemical sense, but are termed basic cations for historical reasons. Within soil minerals, these cations are generally associated with oxides, hydroxides, and carbonates; these anions neutralize H^+ when the cations are released to solution. Moreover, an exchange complex dominated by the basic cations will have a soil solution with a relatively high pH because the exchange complex behaves as a dissociated acid. Acidic cations are H^+ and Al^{3+}, with minor contributions from other hydrolyzing cations, such as iron; exchange complexes dominated by these ions behave as undissociated acids and maintain soil solutions at lower pH values.

A major concern over inputs of acidic deposition to forests is that soil exchange complexes will become dominated by acidic cations, basic cations will be leached from the soil, and soil pH will decline. Predicting the magnitude of these changes is a difficult task that requires a quantitative understanding of the processes regulating the release and retention of all major solutes with the soil and soil solution. For example, Reuss and Johnson (1986) modeled the

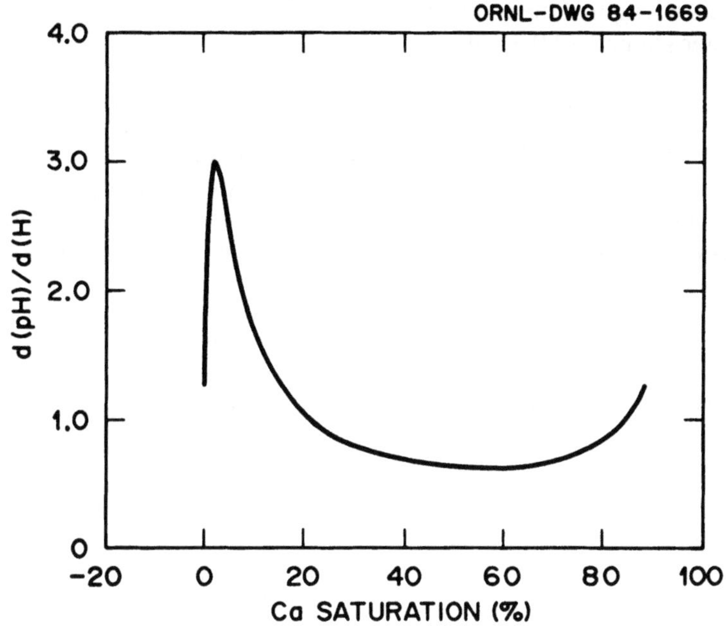

Figure 2.2. The simulated rate of change in soil pH for each H^+ added varies with the degree of calcium saturation of the exchange complex. As with all acid/base systems, the maximum resistance to change in pH occurs in the vicinity of 50% dissociation, or in this case, 50% Ca saturation of the exchange complex. (From Reuss and Johnson 1986.)

rate of change in soil pH as a calcium-dominated exchange complex was incrementally acidified to an aluminum-dominated system by strong acid addition (fig. 2.2). When the calcium concentration of the soil was high (= high base saturation, = highly dissociated exchange complex), the rate of change in pH of the soil solution was low for every increment of acid added. As the soil acidified, slight changes in the calcium pool resulted in more rapid changes in pH for every addition of H^+. Finally, as the exchange complex became dominated by aluminum, the resistance of the system to further change in pH declined. These complex interactions are qualitatively easy to represent in computer simulation models (see chap. 7), but their specific application (and ultimate verification) to real soils is an area of intense research.

Changes in Soil Involving the Dynamics of Weak and Strong Acids

The pH of two soils (or of one soil at different times) may differ because of any combination of four possible factors:

1. The size of the exchange complex.
2. The degree of dissociation of the complex (base saturation).

3. The acid strength of the complex.
4. The ionic strength of the soil solution.

Factor 1—Size of the Exchange Complex

Two soils may differ in pH simply because of differences in the quantity of acids they contain. This characteristic can be illustrated by considering a solution containing acetic acid. If a solution contains 1 mmol/L of acetic acid (CH_3COOH), the H^+ concentration can be calculated from the equilibrium equation:

$$K_{eq} = 1.8 \times 10^{-5} = \frac{[H^+]\,[CH_3COO^-]}{[CH_3COOH]} \tag{15}$$

Let the H^+ concentration equal X, and the acetate concentration will also equal X, and the acetic acid concentration will be [1 mmol/L $- 2X$]. The solution of this equation gives a concentration of H^+ of 1.3×10^{-4} mol/L, or a pH of 3.9. If the solution contained only 0.1 mmol/L of acetic acid, it would have a pH of 4.4.

This simple acid-quantity mechanism has two major implications for forest soils. Soils with high clay (argillic acid) content or a high proportion of clays with high charge density (CEC, such as smectites) will tend to have lower pH values than soils with less caly or with clays of lower charge (such as kaolinites). Of course, other factors may overshadow these trends. In addition, soils that have more organic matter tend to have lower pH because of the greater variable charge CEC. Note that in the acetic acid example a 10-fold change in acid concentration changed the pH by 0.5 units (a 3.2-fold change in H^+ concentration). A 2-fold change in acid concentration would change pH only by 0.15 units. The charge of smectite clays may be 10-fold greater than that of kaolinites, and the clay content of a soil may be 10-fold greater than in another soil. Therefore, differences in clay types and clay quantities should be expected to have a substantial (>0.5 pH unit) effect on soil pH. The organic matter content of soils may also differ by 10-fold or more between soils. However, the clay content of individual soils does not change substantially except over very long time scales, and the organic matter content of soils probably never changes by more than 50% within a century. Therefore, this first mechanism should only account for major pH differences between very different types of soils or across long periods of time.

Factor 2—Degree of Dissociation of the Exchange Complex

The degree of dissociation of the exchange complex is commonly the focus of discussions of impacts of acidic deposition on forest soils. This characteristic can also be illustrated with a simple example with acetic acid. If a 1 mmol/L solution of acetic acid were 25% dissociated, the pH would be 4.2. If it were 75% dissociated, the pH would be 5.2.

Changes in the degree of dissociation of the soil exchange (or acid) complex are usually described by soil scientists as differences in base saturation, and as the illustration in figure 2.2 indicates, reasonable changes in the degree of dissociation can lead to substantial changes in soil-solution pH.

Factor 3—Acid Strength of the Exchange Complex

The changes in the acid strength of the exchange complex is not often considered in analysis of the effects of acidic deposition, but it can be very important from many ecosystem perspectives. The strength of acids is quantified by equilibrium constants (as described earlier); since the concentration of H^+ is often expressed as pH, the same approach is used to produce pK's from K_{eq}'s. The pH of two soils with the same quantity of acids and the same base saturation could differ if the average acid strength of the exchange complex differed. Changes in the quality of the organic matter in a forest soil can have a strong influence on soil acidity. In most cases, such changes occur only over long time periods. However, more rapid changes are possible (see the case study later in this chapter).

This change in the acid strength of an exchange complex has been discussed in the acid deposition literature primarily as shifts in the pH-buffering systems of forest soils undergoing acidification. Ulrich (1987) described six buffer system ranges to cover the range of pH from 8.6 down to less than 3.2. Although these ranges are presented as sequential, some buffering from each process occurs over a much wider range of pH than characterized by Ulrich (1987). Between pH 8.6 and 6.2, the primary reaction buffering pH relative to H^+ inputs is the dissolution of calcium carbonate to form H^+ and bicarbonate. For every 2 mol H^+ added to the system, 1 remains free, and 1 is consumed (and stored or exported) in the formation of bicarbonate. A soil containing calcium carbonate should have a pK above 6; if stronger acids were present to lower the pK, the carbonate would be unstable and dissolve.

The next pH-buffer system is regulated by the weathering reactions of silicate minerals to form silicic acid and often dominates acid/base reactions when soil pH is between about 6.2 and 5.0. Free acidity added to a soil dissolves the silicate minerals, and the H^+ is consumed by the release of basic cations (and formation of silicic acid) from minerals. Again if the pH of the soil decreases below this range, the rate of silicate mineral weathering increases, thus serving to deplete this pool. Rates of weathering of silicate minerals may be rapid enough to neutralize all of the net production of H^+ in some ecosystems (see Clayton 1987). However, some soils lack readily weatherable minerals and high inputs of strong acids may overwhelm this process, resulting in soil acidification beyond the level this pH-buffering system would otherwise maintain.

The third buffer range is regulated by dissociation of the exchange complex. Covering the pH range of about 5.2 to 4.0, this pH buffer system is probably important for most forest soils. Unlike the other buffering mechanisms described by Ulrich, the anionic character of the exchange complex is associated with the soil itself. Whereas the acidity consumed in mineral weathering may be exported from the soil in the form of silicic acid, the acidity removed from soil solution in cation exchange remains within the system. Acidification of the exchange complex can result in lower soil pH (by factor 2), even though added H^+ are largely removed from the soil solution.

Below pH 4.0, aluminum hydroxides begin to dissolve, releasing OH^- to buffer inputs of H^+. Aluminum is the most abundant metallic element in the Earth's crust, so most soils have large amounts of aluminum hydroxides and few non-Histosol forest soils exhaust this buffer system.

Organic acids are important in regulating dynamics in soil acidity, although they are not included in Ulrich's buffering scheme. The pK's of many organic molecules falls in the range of 4.5 to 5.5, but the distribution of pK's can extend from 3.0 to 8.0 or higher.

Factor 4—Ionic Strength of the Soil Solution

The fourth and final mechanism is the ionic strength of the soil solution (termed the *salt effect*). If a salt solution is added to a soil, the cation added will quickly equilibrate with the exchange complex, displacing ions from the exchange complex. If the soil exchange complex is dominated by acidic cations (aluminum), the addition of a salt will lower the pH of the soil solution. This mechanism accounts for how a relatively small input of acid (say, 1 kmol of HNO_3/ha) into a soil already containing very large pools of acid (say, 1000 kmol/ha) can change the soil pH. If addition of nitric acid increases the ionic strength of the soil by twofold, the quantity of H^+ in the solution may double, decreasing pH by 0.3 units. Over the short term, the same effect would result from addition of sodium nitrate salt, or sodium chloride salts—hence the name "*salt effect.*" (Note that this is not the salt effect defined by aquatic chemists; Stumm and Morgan [1981] use the term *salt effect* to describe the decrease in activity coefficients with increases in ionic strength of a solution.) This "salt-effect" is comprised of two separate steps: the displacement of aluminum ions from the exchange complex and the dissociation of H^+ from the hydration sphere of the aluminum. The ionic strength of precipitation in polluted regions is higher than in remote areas, and this feature in itself can lower the pH of soil solution without substantially changing the nature of the exchange complex. Similarly, Binkley et al. (in prep.) found that soil solutions under nitrogen-fixing alders had higher nitrate (and total ionic strength) than solutions under adjacent Douglas-fir stands, and they concluded that the salt effect accounted for about 0.3 units decrease in the pH of the soil solution under alder.

Conservation of Acidity and Alkalinity

Changes in the degree of dissociation of the exchange complex in the soil have implications for the conservation of acidity and alkalinity (*sensu* Stumm and Morgan 1981). When an acid reacts with a base to form a salt, the acid becomes a weak base and the base a weak acid. For example, mixing acetic acid (a weak acid, $pK_a = 4.71$) with sodium bicarbonate (a weak base) yields sodium acetate (a weaker base) and carbonic acid (a weaker acid $pK_a = 6.3$):

$$CH_3COOH + NaHCO_3 \rightarrow CH_3COONa + H_2CO_3 \quad (16)$$

The acidity of the acetic acid is conserved through the transfer of H^+ to the weaker carbonic acid, and the alkalinity of the bicarbonate is conserved by the formation of the weak base acetate.

Alkalinity and acidity are also conserved when equal quantities of a strong acid (such as hydrochloric acid, $pK_a = -3$) react with a weak base (such as sodium bicarbonate):

$$HCl + NaHCO_3 \rightarrow H_2CO_3 + NaCl \tag{17}$$

In this case, it is evident that the acidity is conserved, since carbonic acid is also an acid. It is less evident that alkalinity is also conserved, since sodium chloride is usually called a salt rather than a base. In reality, sodium chloride is a very weak base, as can be shown by reacting it with a very strong acid, such as perchloric acid ($pK_a = -7$):

$$HClO_4 + NaCl \rightarrow HCl + NaClO_4 \tag{18}$$

The same holds true for reactions of strong acids with strong bases, such as hydrochloric acid and sodium hydroxide:

$$HCl + NaOH \rightarrow H_2O + NaCl \tag{19}$$

Water is a very weak acid ($pK_a = 7.0$), and sodium chloride is a very weak base. Fortunately, the reactions of strong acids with strong bases, and weak acids with strong bases, are not important in soils, so the conservation of alkalinity and acidity is usually simple to follow.

In soils, the important reactions involve weak acids with weak bases or strong acids with weak bases. Carbonic acid produced in soils (driven by high CO_2 concentrations) will react with weak bases, such as exchangeable cations and weatherable minerals. Acidity will be transferred to the exchange complex or to silicate minerals, and alkalinity will be transferred to the soil solution as bicarbonate salts. The reactions of strong acids with weak bases are important in acidic deposition, as strong sulfuric and nitric acids also react with the weak bases in the soil. In this case, alkalinity is transferred to the soil solution as a sulfate or nitrate salt, whereas acidity is transferred to the soil.

The importance of this conservation principle to soil chemistry can be illustrated by considering the interaction of carbonic acid with an exchange complex dominated by calcium. As carbonic acid dissociates, H^+ displaces Ca^{2+} from the exchange complex. This process depletes the acidity of the soil solution while increasing the alkalinity (in the form of bicarbonate). Coincident with solution transformation is the acidification of the exchange complex owing to replacement of Ca^{2+} by H^+. As the bicarbonate provided by this process is leached from the soil, alkalinity is transferred to aquatic ecosystems at the expense of acidification of the soil system.

Because acidity and alkalinity are conserved (any change in one gives a proportional change in the other), it is highly useful to characterize the acidity of a solution or soil as the BNC and the alkalinity as the ANC (see van Breemen et al. 1983). A soil with a large exchange complex dominated by acidic cations (such as aluminum) will neutralize large amounts of a weak base (such as calcium carbonate) with only small increases in soil pH. A soil with small pools of acidic cations would have lower BNC and would exhibit larger increases in pH following additions of a strong base. The ANC of the soil is important with regard to effects of acidic deposition. A soil with an exchange complex dominated by basic cations has a large reserve of alkalinity, or ANC, residing in the basic anion (not the basic cation, which is really a salt cation). A base-rich (high ANC) soil is better buffered against changes in pH than a soil low in ANC.

Deposition of strong acids (such as nitric acid) may acidify the soil by transferring acidity to the exchange complex (decreasing BNC) and alkalinity from the exchange complex to soil solutions (increasing ANC). Addition of weak acids in soil organic matter represents an increase in acidity (BNC), but any dissociation of these acids (to become variable charge CEC sites) results in production of a proportional increase in alkalinity (ANC) of the exchange complex and increase in acidity (BNC) of the soil solution.

Sulfur, Nitrogen, and H^+ Budgets

The discussion of soil acidity presented earlier described changes in the soil solution and exchange complex. Such changes are mediated by processes in the ecosystem that generate or consume H^+ and those that generate or consume organic acids stored in the soil. The total flux of H^+ in ecosystems is enormous, but fortunately only the net fluxes are important at the scale of ecosystem acidification (see Binkley and Richter 1987).

Sources of H^+

The major sources of H^+ in forest ecosystems are:

1. Atmospheric deposition.
2. Cation accumulation in organic matter (living and dead).
3. Formation and dissociation of carbonic acid in the soil.
4. Unbalanced oxidations of reduced compounds.

Atmospheric Deposition

In the southern United States, annual rates of deposition of H^+ probably range from 0.5 to 1.5 kmol/ha annually (see chap. 3). However, the net effect is often determined by the fate of the anion associated with the acidity. Both sulfur and nitrogen are used in plant proteins in the reduced form. Reducing sulfate and nitrate consumes acidity. Therefore, assimilation of sulfuric and nitric acids by the activity of microbes and plants neutralizes acidity. As discussed in chapter 3, the deposition of H^+ to a chestnut oak/white oak forest in Tennessee was about 1.6 kmol/ha annually; retention of the anions within the forest neutralized two-thirds of this acidity (see Binkley and Richter 1987). Southern forests are typically N deficient, and the acidity associated with nitrate deposition is essentially neutralized through plant uptake. Sulfur uptake is less likely to occur, and the H^+ associated with sulfate may have an opportunity to acidify the ecosystem.

Element Accumulation in Biomass

Cation nutrients, such as Ca^{2+} and K^+, do not undergo oxidation and reduction reactions; they retain ionic character. Therefore, plants must have negative charges to balance these cations, and these negative charges are produced by formation of bicarbonate and organic anions. In order to maintain electroneutrality, H^+ must be extruded from the roots in a quantity equivalent to the net accumulation of cations. If biomass accumulates, the storage of cations results in a proportional input of H^+ to the soil. Conversely, when biomass decomposes, the process is reversed and acidity is consumed (fig. 2.3).

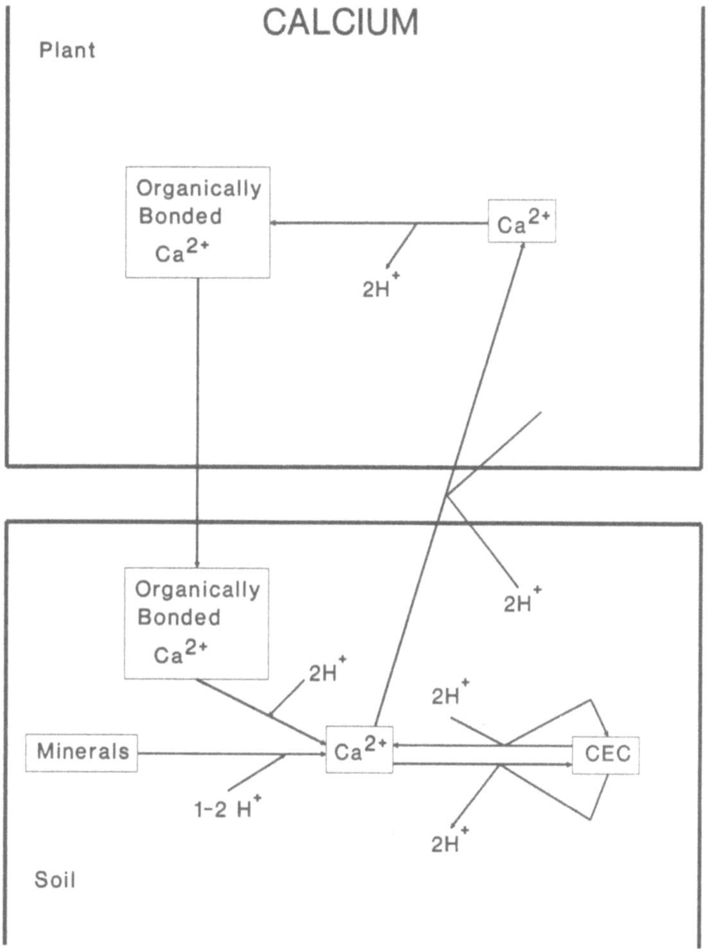

Figure 2.3. The H^+ components of the calcium cycle. Weathering consumes H^+, ionic Ca^{2+} may displace H^+ from exchange sites or be taken up (in exchange for H^+) by plant. Plant uptake of Ca^{2+} released from organic matter has no effect on H^+, but uptake from exchange sites represents H^+ release into the soil. (From Binkley and Richter 1987, reprinted by permission of Academic Press.)

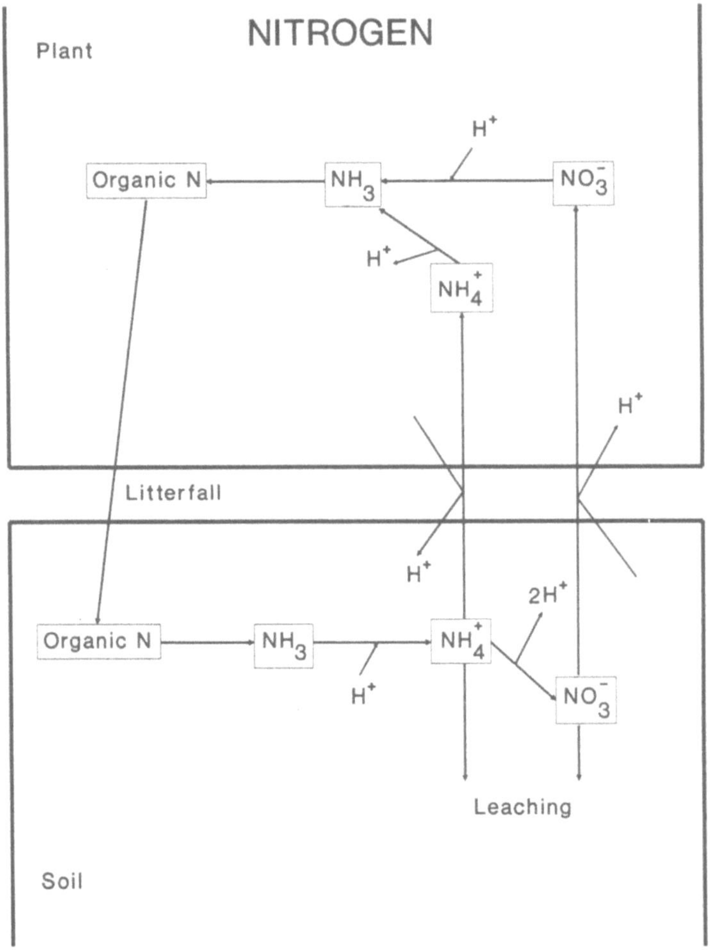

Figure 2.4. Oxidation and reduction reactions in the N cycle generate and consume H^+, but the transfer of organic N from soil organic matter to vegetation has no net effect. Major net fluxes of H^+ are associated only with inorganic forms of N entering or leaving the ecosystem. (From Binkley and Richter 1987, reprinted by permission of Academic Press.)

Anion uptake has the opposite effect of cation accumulation, causing a net flow of bicarbonate out of the plant (or H^+ into the plant). However, the assimilation of sulfate and nitrate involve reductions that consume H^+ (discussed later). The loss of alkalinity (increase in acidity) of the plants is neutralized by the processing of the anions into organic compounds. Furthermore, if the nitrate and sulfate were formed within the soil by oxidation of reduced compounds, the loss of acidity from the soil associated with uptake of these anions would precisely balance an earlier input of acidity (see figs. 2.4, 2.5). For these reasons, the cycles of N and S are balanced with respect to H^+ flux and are important components of H^+ budgets only when additions and losses from ecosystems occur.

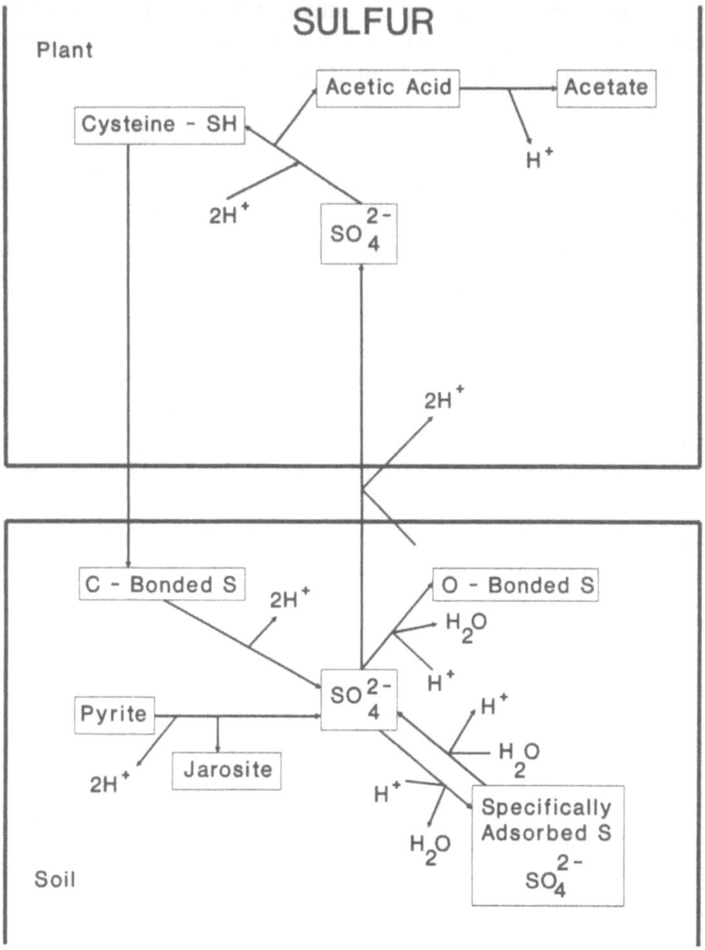

Figure 2.5. The sulfur cycle combines the complexities of the N and P cycles. Major net fluxes of H^+ are associated with inputs and outputs of inorganic ions and with sulfate sorption. (From Binkley and Richter 1987, reprinted by permission of Academic Press.)

The effect of phosphate uptake is variable, but fortunately small enough to be negligible at an ecosystem scale. Some chloride is taken up by plants and would require uptake of H^+ or release of HCO_3^-; but again, the quantities assimilated are small.

Therefore, the net H^+ flux associated with nutrient uptake and storage in biomass is usually gauged simply as the sum of the cations. Ammonium (NH_4^+) that is recycled within the system is excluded from this sum since it releases one H^+ when assimilated. It is also important to realize that some of the cations accumulated in living biomass may have been recycled from detritus rather than have come from the exchange complex or some other source; therefore, the actual rate of H^+ production associated with cation uptake and storage may be less than the amount estimated in live biomass. On the other land, soil organic matter pools may also increase over time and the net flux could also be greater than the aboveground increment in biomass would indicate.

Formation and Dissociation of Carbonic Acid

Respiration by roots and microbes in the soil produces large amounts of carbon dioxide, and relatively slow rates of diffusion of gases out of the soil causes the concentrations of carbon dioxide to accumulate from ten to a hundred times the value for the atmosphere at large. Recall the equation for formation of carbonic acid and LeChatelier's principle: increased carbon dioxide concentrations increase carbonic acid production. The amount of H^+ generated in forest soils by this mechanism varies with the carbon dioxide concentration and soil pH, but it commonly falls within the range of 0.5 to 1.0 kmol/ha annually. For comparison, 1000 mm of precipitation with a pH of 4.0 contains 1 kmol/ha of H^+. Inputs from carbonic acid formed in soil are higher in soils with high pH because more of the acid dissociates. This trend also indicates that export of alkalinity (bicarbonate) to aquatic ecosystems is greatest from soils of high pH and that reductions in soil pH can reduce alkalinity transport to lakes.

Unbalanced Oxidation of Reduced Compounds

An understanding of oxidation and reduction is critical for interpretation of H^+ production in soil systems. Oxidation involves transfer of high-energy electrons to an electron acceptor, and reduction is the process of gaining electrons. This process is coupled to soil acidity through the Lewis definition of acidity. The classic view of acids as H^+ and bases as OH^- (the Arrhenius concept) or of acids as H^+ donors and bases as H^+ acceptors (the Bronsted/Lowry concept) was expanded in the 1920s by Lewis. He defined acids as acceptors of pairs of electrons and bases as donors of pairs of electrons. The Lewis concept includes classic acid/base reactions, such as hydrochloric acid reacting with sodium hydroxide:

$$HCl + NaOH \rightarrow H_2O + NaCl \tag{20}$$

Here hydrochloric acid accepts electrons from sodium hydroxide, coincident with the transfer of a proton to sodium hydroxide.

The value of the Lewis concept becomes evident when considering acid/base reactions that involve oxidation and reduction. For example, when ammonia is oxidized by molecular oxygen, a strong acid (by the Bronsted/Lowry concept) is produced:

$$NH_3 + 2O_2 \rightarrow HNO_3 + H_2O + \text{energy} \tag{21}$$

Ammonia donates electrons and is a Lewis base, whereas oxygen accepts electrons and is a Lewis acid. In keeping with the conservation of acidity and alkalinity, the acidity of the Lewis acid (oxygen) is transfered to H^+ in the nitric acid, whereas the alkalinity resides in the nitrate.

When inorganic N is produced from the decomposition of organic matter (called mineralization) in forest soils, the initial byproduct is ammonia (fig. 2.4). When soil pH is below about 9 (the pK_a of the NH_4^+/NH_3 system is 9.3), NH_3 consumes H^+ from the soil solution to produce the weak acid, ammonium (NH_4^+). The alkalinity associated with NH_4^+ is , in turn, transferred to the soil solution or perhaps to the exchange complex. If a plant assimilates NH_4^+, it releases an equivalent quantity of H^+ into the soil, which neutralizes the alkalinity in the soil and leaves the soil balanced with respect to H^+. The plant has gained alkalinity through the loss of the H^+, but this is neutralized when the ammounium is converted into a protein. The net effect of this process is that all generation and consumption of H^+ is balanced with no net change in the ecosystem or in any compartment of the ecosystem.

When ammonium is oxidized (called nitrification) by autotrophic and heterotrophic microbes (that obtain energy from the reaction), the acidity associated with ammonium is released, and the Lewis acidity of the oxygen is transferred to nitric acid:

$$NH_4^+ + 2O_2 \rightarrow HNO_3 + H^+ + H_2O + \text{energy} \quad (22)$$

Recall that production of the ammonium originally conveyed alkalinity to the soil, so one of the two moles of H^+ produced from nitrification balances the consumption of H^+ associated with the mineralization of organic N, leaving a net production of 1 mole of H^+. The complete dissociation of the strong nitric leaves the "acidity" of the H^+ more apparent than the "alkalinity" of the nitrate. However, if plants take up the nitrate and assimilate it, the nitrate will be reduced and H^+ will be consumed:

$$9H^+ + NO_3^- + 8e^- \rightarrow NH_3 + 3H_2O \quad (23)$$

$$(\text{and } NH_3 + \text{organic} \rightarrow \text{protein})$$

The same net effect of acidity consumption occurs if the nitrate is used as a terminal electron acceptor (called denitrification, or dissimilatory nitrate reduction):

$$2H^+ + 2NO_3^- \ (+\ 10\ H^+ + 10e^-) \rightarrow N_2 + 6H_2O \quad (24)$$

Again the complete cycle of nitrogen from organic nitrogen in the soil, through the transformations to ammonium and nitrate, back into organic N in plants, involves no net production of acidity or alkalinity. However, this cycle is interrupted if nitrate is leached from the soil. When nitrate is exported from a soil system, the acidity associated with nitrification typically remains in the soil (nitrate's negative charge is usually balanced by base cations), and alkalinity associated with nitrate is exported. Nitrate leaching in most forests is negligible, but in some cases, such as the disturbance associated with harvesting northern hardwood forests or in ecosystems with nitrogen-fixing black locust or red alder, nitrate leaching can be substantial, generating up to 1 to 3 kmol/ha of H^+

annually (Likens et al. 1970, Van Miegroet and Cole 1984). If the nitrate exported is subsequently reduced in aquatic ecosystems, either through biotic uptake or denitrification, alkalinity is transferred to the water.

The sulfur cycle (fig. 2.5) resembles the nitrogen cycle, with a few extra complications.

Processes Consuming H^+

Some of the processes consuming H^+ in ecosystems were described earlier; the major processes include:

1. Release of basic cations through decomposition or fire.
2. Specific anion adsorption.
3. Mineral weathering.
4. Unbalanced reduction of oxidized compounds.

Release of Basic Cations

Just as accumulation of basic cations in organic matter transfers acidity to soils (and generates alkalinity in plants), the release of these cations produces alkalinity in the soil, neutralizing the earlier production of acidity. For example, if an organic molecule containing calcium is oxidized, H^+ must be consumed in the production of carbon dioxide and water:

$$\text{COO-Ca-OOC} + 2H^+ + \tfrac{1}{2}O_2 \rightarrow Ca^{2+} + 2CO_2 + H_2O \qquad (25)$$

The reaction is the same whether the oxidation of the organic matter occurs during decomposition by microbes or in a fire and explains why ash from fires is alkaline.

After a forest is harvested, the rate of organic matter decomposition often exceeds the rate of production of new biomass; therefore, H^+ consumption in decomposition should exceed that produced by cation accumulation in biomass. A wildfire would have the same result, only much more quickly! For this reason, soil pH may increase after harvest, unless nitrication coupled with nitrate leaching overrides this mechanism.

Specific Anion Adsorption

Sulfate and phosphate may be specifically adsorbed (enter into the hydration sphere) by aluminum and iron sesquioxides (Tabatabai 1987) in the soil:

$$2\text{R-OH} + SO_4^{2-} + 2H^+ \rightarrow (\text{R-OH})_2SO_4 + H_2O \qquad (26)$$

This adsorption transfers alkalinity associated with sulfate to the soil solution through consumption of $2H^+$. This process is a major factor in sulfate retention in soils experiencing increased atmospheric inputs, and it is a major determinant of whether deposited H^+ will be neutralized or have an opportunity to acidify the soil. Of course, the acidity is actually retained in the soil at the sesquioxide surface, and subsequent desorption would restore the alkalinity to the sulfate ion and the acidity to the soil solution.

If the rate of sulfate deposition to soils decreased, the desorption of sulfate could become an important process because the release of the associated acidity might delay the recovery of soils by decades (see fig. 6.2). The reversibility of sulfate adsorption is an important research issue, but it is poorly understood at present (Singh 1984).

D.W. Johnson et al. (1980) surveyed concentrations of adsorbed sulfate and factors regulating these concentrations in several types of soils. They found that Spodosols formed on glaciated parent materials generally contained low concentrations of adsorbed sulfate, whereas some Inceptisols formed on glaciated parent materials and other soils from nonglaciated regions contained higher concentrations. Free iron and aluminum (nonsilicate bound) in soil effectively adsorb sulfate (Chao et al. 1964, Harward and Reisenaur 1966). D.W. Johnson et al. (1980) also reported that the free-iron content of soils correlated with adsorbed soil sulfate, but aluminum did not. Fuller et al. (1985), however, observed that adsorbed sulfate correlated strongly with free aluminum in a New Hampshire soil.

The concentrations and transport of counterions are critical in regulating aluminum mobility in acid-sensitive regions. Some calculations made with the ALCHEMI model (Schecher and Driscoll 1987; see figs. 2.6, 2.7) show the effects of incremental concentrations of sulfuric acid on solutions initially with 100 $umol_c/L$ of bicarbonate and in equilibrium with $Al(OH)_3$ at two different levels of sulfate adsorption. A low concentration of surface sites (5.6 m^2/L) represents a system with low sulfate adsorption capacity (fig. 2.6), such as a glaciated Spodosol. The higher concentration of surface sites (560 m^2/L) facilitates sulfate adsorption (fig. 2.7), and represents older, nonglaciated soils high in free iron and aluminum (such as Ultisols).

At low concentration of added sulfuric acid, pH values are near-neutral (7.3 to 7.6), aluminum concentrations are low (<1.0 μmol/L) and all sulfate is in the aqueous phase. As sulfuric acid is added, pH decreases. As values approach the pH of minimum solubility of $Al(OH)_3$, aluminum concentrations decrease. With further acid addition, pH decreases below the minimum solubility level for $Al(OH)_3$ and aluminum concentrations increase markedly. In the presence of low concentrations of surface adsorption sites, sulfate inputs remain largely in solution. Relatively small inputs of acid are required to acidify the solution and to increase aluminum concentrations. Such conditions are similar to those reported for acidic waters in the northeastern United States (N.M. Johnson et al. 1981) and Scandinavia (Dickson 1978). At low pH values and high sulfate inputs, the solubility of the aluminum sulfate mineral, jurbanite [$Al(OH)SO_4 \cdot 5H_2O$] is exceeded. Additional sulfate inputs are precipitated as jurbanite, and this reaction buffers the system against further decreases in pH and increases in aluminum concentrations.

In contrast to the situation for the low-adsorption simulation, the high-adsorption-capacity soil retained much of the added sulfate under acidic conditions. In fact, sulfate adsorption increases as pH decreases because of pro-

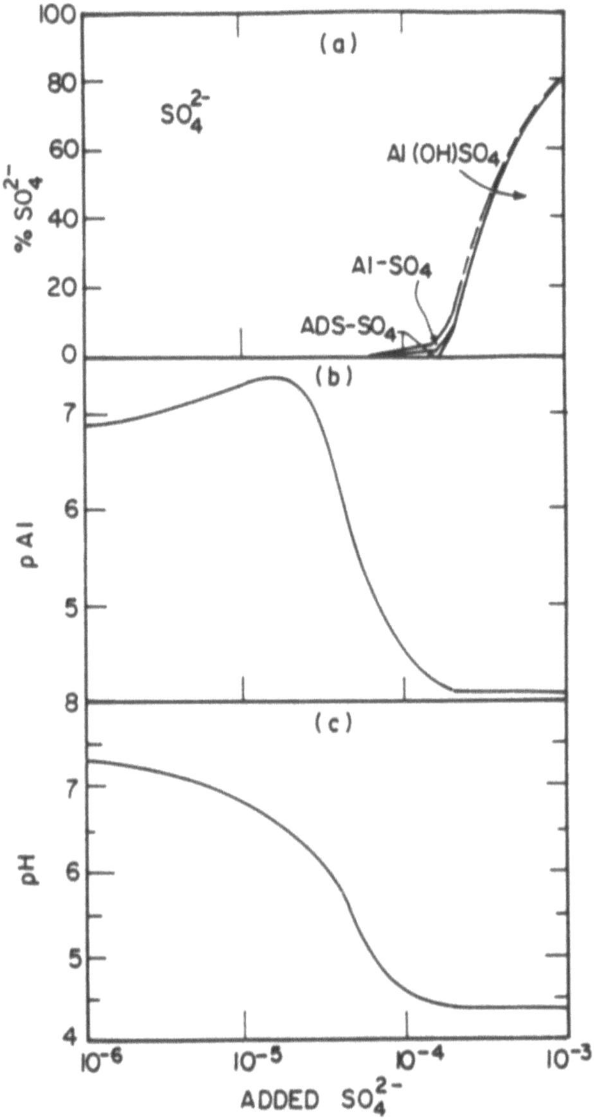

Figure 2.6. Simulations with the ALCHEMI model examining changes from initial conditions of 100 μmol HCO_3^- in equilibrium with $Al(OH)_3$. With low sulfate adsorption capacity, most added sulfate remains in solution until high concentrations favor precipitation of aluminum sulfate compounds (see text). (Reprinted from Schecher and Driscoll 1987, by courtesy of Marcel Dekker, Inc.)

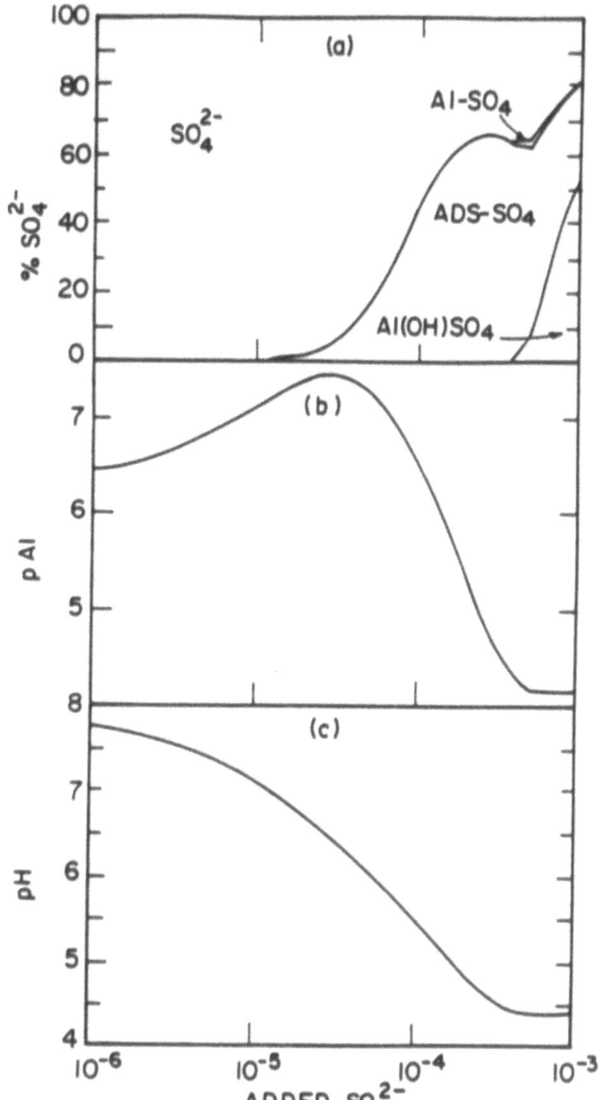

Figure 2.7. Hypothetical titration with sulfate of a soil system with high sulfate adsorption capacity (see fig. 2.6 and text). (Reprinted from Schecher and Driscoll 1987, by courtesy of Marcel Dekker, Inc.)

tonation of adsorption sites (Davis and Leckie 1980). In the presence of high adsorption capacity, large concentrations of added sulfate are needed to lower pH and mobilize aluminum. Jurbanite also fails to form if adsorption keeps the solution concentrations of sulfate too low.

These patterns illustrate how proposed mechanisms that regulate solution-aluminum concentrations may differ among regions; the solubility of $Al(OH)_3$ may regulate aluminum concentrations in the northeastern United States (N.M. Johnson et al. 1981), whereas jurbanite solubility may regulate concentrations in Europe (Nilsson and Bergkvist 1983). The apparent discrepancies among regions may simply reflect differences in rates of sulfate deposition and solution concentrations in the two regions (van Breemen et al. 1984).

These simulations and discussions show that sulfate retention is a critical factor that regulates acidification of soils and drainage waters. Soils that strongly retain sulfate are unlikely to experience significant acidification or mobilization of aluminum.

Mineral Weathering

The weathering of minerals can consume or release H^+, but in most situations, H^+ consumption predominates (see Binkley and Richter 1987). The weathering of carbonaceous sedimentary rocks can be illustrated as the dissolution of limestone:

$$CaCO_3 + 2H^+ \rightarrow Ca^{2+} + H_2CO_3 \rightarrow Ca^{2+} + H^+ + HCO_3^- \quad (27)$$

Under acidic conditions ($pH < 5$), the carbonic acid will largely remain undissociated, and 2 moles of H^+ will be consumed for every mol of Ca^{2+} released. At higher pH, only 1 H^+ will be consumed.

The weathering of aluminosilicate minerals consumes a mole of H^+ per mole of charge of cation released. The weathering of calcium feldspar is representative:

$$CaAl_2SiO_8 + 8H^+ \rightarrow Ca^{2+} + 2Al^{3+} + 2H_4SiO_4 \quad (28)$$

Aluminum may then hydrate and, depending on the pH of the soil solution, it releases a variable number of H^+ as a result of the extent of the reaction:

$$2Al^{3+} + 12H_2O \rightarrow 2Al(H_2O)_4(OH)_2^+ + 4H^+ \quad (29)$$

The orthosilicic acid [H_4SiO_4] formed is a very weak acid that remains essentially undissociated if the pH is below about 9. It is also possible for the aluminum and silicic acid to form the secondary mineral, kaolinite:

$$2Al^{3+} + 2H_4SiO_4 \rightarrow Al_2Si_2O_5(OH)_4 + 6H^+ \quad (30)$$

The net effect is the transfer of acidity from a strong acid (H^+ in solution) to a weak acid (silicic acid), with the associated release of calcium and aluminum.

The acidity is then removed from the system through the leaching of the silicic acid, or it is stored in the form of secondary clay minerals.

Unbalanced Reduction of Oxidized Compounds

If a compound is oxidized within a system, transported, and then subsequently reduced in a different system, the consumption of H^+ will not be coupled with the generation of H^+ within the initial system. For example, if potassium nitrate fertilizer is added to a forest, uptake and reduction of the nitrate will consume H^+, as would loss of the nitrate through denitrification. This latter mechanism is generally unimportant in forest ecosystems because most forests experience aerobic conditions (where dissimilatory reduction is minimal), and they receive little input of oxidized compounds except those associated with their own H^+ (such as sulfuric acid rain). The pattern is quite different for aquatic systems, which often have anaerobic (reducing) sediments and experience inputs of oxidized compounds (such as nitrate) that are associated with basic cations.

Loblolly Pine Case Studies

The linkage between acidic deposition, soil acidity, and H^+ budgets can be illustrated by two examples from intensive study sites. These study sites were chosen by the original investigators as generally representative of common sites within the region, but given the great diversity of ecosystems and soils across the southeastern United States, these sites can only serve as illustrations and not as generalizable examples. A H^+ budget for a 20-yr-old plantation of loblolly pine (planted in an abandoned agricultural field) in the Duke Forest in North Carolina is being prepared as part of the Integrated Forest Study coordinated through Oak Ridge National Laboratory and funded by the Electric Power Research Institute. Some information on long-term changes in soil chemistry is available from a similar loblolly pine plantation in South Carolina, where C.G. Wells sampled soils at yr 5 and 25 of the plantation's development.

The major net fluxes of H^+ for the plantation in North Carolina are obtained from unpublished data of K. Knoerr and D. Binkley of Duke University for a 20-yr-old forest on a clayey, kaolinitic, thermic, Typic Hapludult. The wet + dry deposition of free H^+ was about 0.9 kmol/ha annually (table 2.1). The production of H^+ from formation and dissociation of carbonic acid was about 0.3 kmol/ha annually, estimated from the amount of bicarbonate leaching from the soil. The annual accumulation of cation nutrients in this rapidly aggrading forest was about 0.6 kmol/ha; about half of which came from cations deposited from the atmosphere and half from the exchange complex. Sulfate outputs exceeded wet + dry inputs (probably because of the leaching of S added as part of P fertilizers during agricultural cropping), with a H^+ generation of 0.4 kmol/ha annually. Therefore, the total production of H^+ summed to 2.2 kmol/ha annually, and the leaching loss of base cations was 1.5 $kmol_c$/ha.

The net consumption of H^+ was much less. Because biomass was strongly aggrading, there was no net release of basic cations from decomposing biomass.

Table 2.1. The major H^+ budget components for the 20-yr-old loblolly pine forest in Duke Forest, North Carolina

	kmol H^+/ha annually
Inputs	
Deposition	0.9
Cation accumulation in biomass	0.6
Carbonic acid dissociation	0.3
Unbalanced oxidation of S compounds	0.4
Sum	2.2
Outputs	
Net release of cations from biomass	0.0
Specific anion adsorption of S	0.0
Mineral weathering	0.0?
Unbalanced reduction of oxidized compounds	0.2
Sum	0.2
Net annual change in ecosystem acidity pools	+2.0[a]

[a]Unmeasured fluxes: Changes in soil organic matters; weathering (probably 0?); changes in acid strength of exchange complex.
Preliminary data from K. Knoerr and D. Binkley.

No net adsorption of sulfate occured. The rate of weathering was not measured, but it was probably slight in this very old soil (a sandy loam, clayey, kaolinitic thermic Typic Hapludult). Therefore, the only substantial consumption of H^+ was probably due to assimilation of nitrate deposited from the atmosphere, about 0.2 kmol/ha annually.

The net H^+ balance for the North Carolina pine forest gives an annual increase of about 2.0 kmol H^+/ha. About 0.9 kmol H^+/ha were deposited from the atmosphere, and assimilation of the accompanying nitrate anion consumed 0.2 kmol/ha; therefore, the net contribution of atmospheric H^+ input was 0.7 kmol/ha, or about one-third of the net increase of H^+ in the ecosystem.

This rate of net H^+ production has important implications. The changes in soil chemistry from the South Carolina pine forest illustrate changes to be expected in a poorly buffered soil (a clayey, mixed thermic Aquic Hapludult). Binkley et al. (1989) analyzed soil samples from a loblolly pine stand collected by C. Wells at yr 5 and yr 25 for changes in pH, exchangeable cations, acidity (BNC), and alkalinity (ANC). Soil chemistry changed dramatically during 20 yr of stand development, with a decline in pH of 0.3 to 0.8 units in all horizons. The ANC declined marginally, whereas the BNC increased markedly. The content of exchangeable bases decreased by 20% to 80%. The expected pH in 1982 of single factors was set equal to the 1962 value are listed in table 2.2. For example, if the exchangeable bases remained at the 1962 value, but the acid strength of the exchange complex and the BNC were as in 1982, the pH would be expected to be 5.26 in the 0 to 7.5 cm soil rather than the actual 3.93. This expected pH is actually higher than the 1962 (4.52), which can be true only if the average acid

Table 2.2. Expected soil pH in 1982 if single factors set equal to 1962

	pH_{KCl}		Expected 1982 pH if Single Factor Set Equal to 1962		
Depth (cm)	1962	1982	BNC	Bases	Acid Strength
0–7.5	4.52	3.93	4.37	5.26	3.71
7.5–15	4.73	4.26	4.69	5.23	4.11
15–*B*	4.75	4.36	4.71	4.89	4.29
B	4.49	4.17	4.31	4.31	4.29

From data in Binkley et al. (1989).

strength of the exchange complex decreased (meaning the acid strength was weaker in 1982 than in 1962). The change in BNC also appeared important in reducing the pH over 1962 values.

Soil solutions were not sampled in the South Carolina site, but simulations with the Modeling Acidification of Groundwater in Catchments (MAGIC) model provide some insight. The model does not account for any changes that may have occurred in the selectivity coefficients for cation exchange or in the acid strength of the exchange complex, thus the predicted chemistry is only a first approximation. The simulation produced reasonable estimates of changes in base saturation for this soil over the 20-yr period (fig. 2.8), so the predicted changes in soil-solution chemistry may also be reasonable. The MAGIC model estimated that the soil solution in 1962 may have had an aluminum concentration of 0.3 $\mu mol_c/L$, and an alkalinity of 125 $\mu mol_c/L$. By 1982, the soil solution may have had about 3.0 $\mu mol_c/L$ of aluminum, and 42 $\mu mol_c/L$ of alkalinity. These

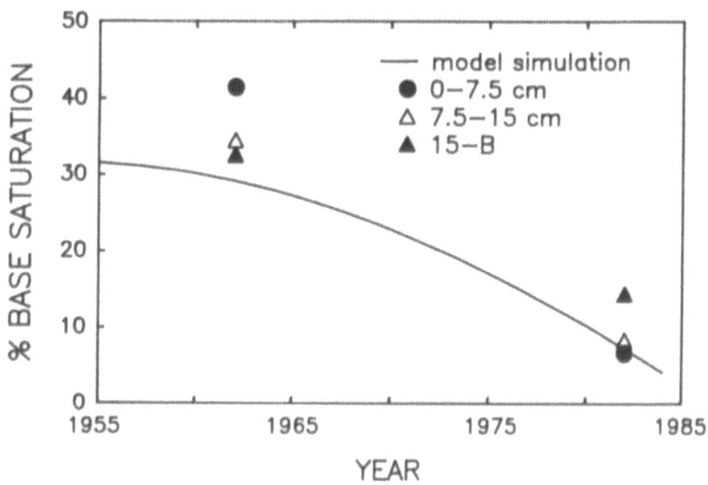

Figure 2.8. Simulations with the MAGIC model generally followed the trend of decreasing base saturation over 20 yr for the South Carolina soil.

simulations demonstrate that soil-solution chemistry may be very sensitive to changes in the exchange complex and that these changes may be driven largely by natural stand development processes. A tenfold increase in solution aluminum concentrations in just 20 yr could have important on-site biologic implications, while a $\frac{2}{3}$ reduction in alkalinity of exported waters could affect the biogeochemistry of aquatic ecosystems.

This soil was very poorly buffered with respect to changes in pool sizes. The soil at age 5 had only about 150 $kmol_c$/ha of basic cations, and the average annual depletion of the basic cations over 20 yr was 2.2 $kmol_c$/ha, which matched the rate of increase in exchangeable aluminum. If the accumulation of nutrient cations in biomass matched the North Carolina pine forest (0.6 $kmol_c$/ha annually), the leaching of basic cations was probably about 1.6 $kmol_c$/ha, similar to the estimate for the North Carolina forest. The North Carolina soil had more than twice the content of exchangeable basic cations and cation exchange capacity, however, so its pH should change more slowly than the South Carolina soil did.

The major implications of these two case studies are:

1. Atmospheric deposition accounted for about 40% of the H^+ input to the forest.
2. Assimilation of the nitrate deposited from the atmosphere neutralized about 20% of the deposited H^+, leaving a net contribution from atmospheric deposition of only 0.7 kmol H^+/ha annually.
3. The rate of increase in H^+ of the soils should be on the order of 2 to 3 kmol H^+/ha annually for rapidly aggrading pine stands. This value compares favorably with the range reported by van Breemen and Mulder (1987) of 0.5 to 3.2 kmol H^+/ha, based on the age of soils and their accumulated BNCs.
4. This rate of H^+ accumulation should be expected to alter the pH of poorly buffered soils; however, shifts in other factors (such as acid strength of the exchange complex) may also affect pH changes.

The complex interactions of acidic deposition and forest biogeochemistry require that evaluations of the likely impacts of acidic deposition must be examined at an ecosystem scale. Computer simulation models are needed to keep track of these interactions, but these models may not adequately account for all the interactions. For example, the models currently available do not simulate changes in forest productivity as a result of changing nutrient availability, and they do not include changes in the acid strength of the exchange complex. However, such models are the best available method for synthesizing current understanding and for determining the research areas with the highest potential importance.

3. Magnitudes and Patterns of Nitrogen and Sulfur Deposition in the South

Precipitation in the South currently has a pH of about 4.4 to 4.6, and about 75% to 80% of the free acidity is associated with sulfuric acid, whereas the remainder is derived from nitric acid. Carbonic acid is present in the rain, but in this low pH range, it is essentially fully protonated. Dry deposition adds a largely unquantified amount of chemicals to forests; dry deposition probably exceeds wet deposition in some areas. In this chapter, the sources of sulfur and nitrogen compounds in the atmosphere and the pathways of deposition and the state of knowledge about historic and current deposition rates are summarized.

Sources of Acids in the Atmosphere

As mentioned in chapter 2, carbonic acid is a natural component of water in the atmosphere because of the dissolution of carbon dioxide in water. Sulfate and nitrate are natural components of the atmosphere even in unpolluted regions, where the concentrations of these ions are about 10% to 15% of the values currently reported in the South (Galloway et al. 1984). Some of the natural sulfate in precipitation is derived from entrainment of particles into the air from the oceans and from terrestrial dust (for a global perspective, see Mooney et al. 1987). As such, the sulfate would be a salt (associated with base cations) and not acidic. Biologic processes also transfer some sulfur to the atmosphere, as reduced H_2S and in organic forms, such as COS and CS_2. The rate of biologic

Table 3.1. Comparisons of best estimates of sulfur sources on a global scale (millions of Mg/yr)

	Minimum	Medium	Maximum
Natural fluxes			
Volanoes	5	10	30
Sea spray	40	45	60
Biogenic S	70	140	175
Sum	115	195	265
Human-related fluxes	82	98	112

After Smil 1985.

production is uncertain, but it is probably about 6 mol/ha annually on a global basis (Adams et al. 1980), which is about the rate for the South (National Acid Precipitation Assessment Program [NAPAP 1985]). This value is small compared to current emission rates from human activities in the South (discussed later), but on a global basis, natural sources probably account for more than half of the atmospheric S (Bubenick 1984). Some S also enters the atmosphere from volcanic activity (table 3.1), but on a global scale, volcanoes probably contribute only a small percentage of the total S in the atmosphere (Smil 1985).

Some nitrate in rain is produced from the oxidation of N_2 associated with lightning, but it is largely due to emissions from soils, combustion, and the oceans. Natural sources probably exceed human-related sources on a global basis (Bubenick 1984), but the magnitudes are not well established.

On a more restricted regional scale, anthropogenic sources can be much greater than natural sources; acidic deposition is an important problem for industrialized regions. Elevated concentrations of sulfate and nitrate in rain in industrialized regions comes from emissions of sulfur and nitrogen oxides that are converted to acids in the atmosphere (NAPAP 1985). The major sources of sulfur oxides are point-source industries, such as power plants that burn high-sulfur coal (producing about two-thirds of the S emissions across the United States). Nitrogen oxides come from a broader array of sources; when any substance is burned or oxidized in the presence of N_2 (which comprises 78% of the atmosphere), some nitrogen oxides are formed. Therefore, major sources include not only power plants (one-third) and other industrial sources (one-fourth), but also motor vehicles (one-third).

The atmosphere also contains chemicals that act as bases. For example, emission of ammonia from animal wastes or other sources conveys alkalinity to rainfall by consuming H^+ to produce ammonium. This process might seem to be an important mechanism for neutralizing some of the acidity in acidic deposition, but recall from chapter 2 that acidity and alkalinity are conserved. Therefore, the ammonium retains the acidity initially derived from sulfuric or nitric acid. If the ammonium is deposited in an ecosystem and is assimilated by plants, H^+ is released from the roots and is available to acidify the soil and drainage waters. Ammonia is also a Lewis base, but in the presence of oxygen (a strong

Lewis acid), microbes generate strong nitric acid (and also gain energy) by catalyzing the reaction of ammonia with oxygen. This source of acidification is an especially important problem in parts of Europe (such as the Netherlands), where deposition rates of ammonium may be as high as several $kmol_c$/ha annually (van Breemen and Mulder 1987).

The atmosphere also contains substantial quantities of basic cations, such as Ca^{2+}. Rainfall with a high concentration of basic cations tends to be less acidic because these cations are largely associated with true bases, such as bicarbonate. However, if basic cations deposited on forest ecosystems are accoumulated into biomass, a net production of acidity in the soil would occur. If the forest were harvested, the net production would be permanent, but if it decomposed in place or burned, the acidification effect would be reversed (see chap. 2).

Estimates of the rates of entrainment of bases and basic cations into the atmosphere are not available, but some deductions can be made. For example, most sulfate and nitrate in precipitation probably originated as acids, yet the H^+ content of rain is usually less (on a charge basis) than the sum of sulfate plus nitrate (see table 3.2).

Anthropogenic Emissions of Sulfur and Nitrogen Oxides

The record of rates of emission of S and N have been reconstructed back into the last century for the eastern United States (Husar 1986). In 1900, about 0.15 kmol of sulfur were emitted for every hectare in the South (fig. 3.1), while emissions

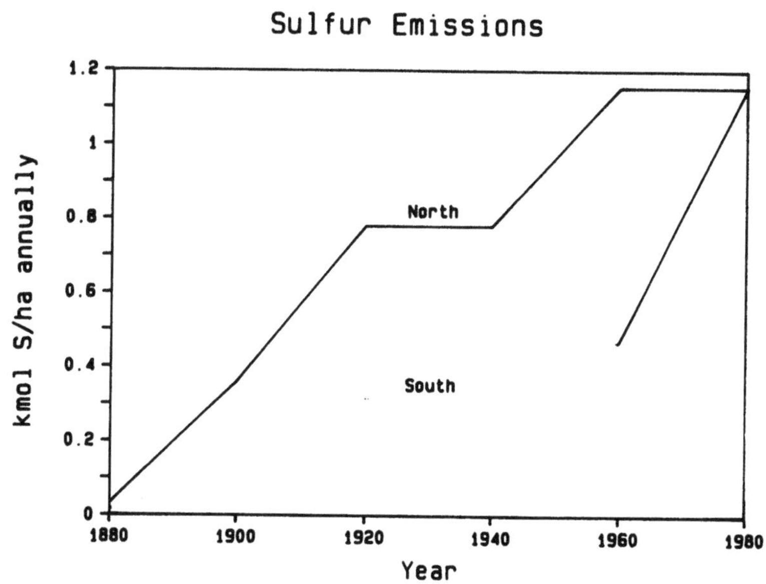

Figure 3.1. Sulfur emission for the eastern United States. (After Husar 1986, reprinted by permission of the National Academy Press.)

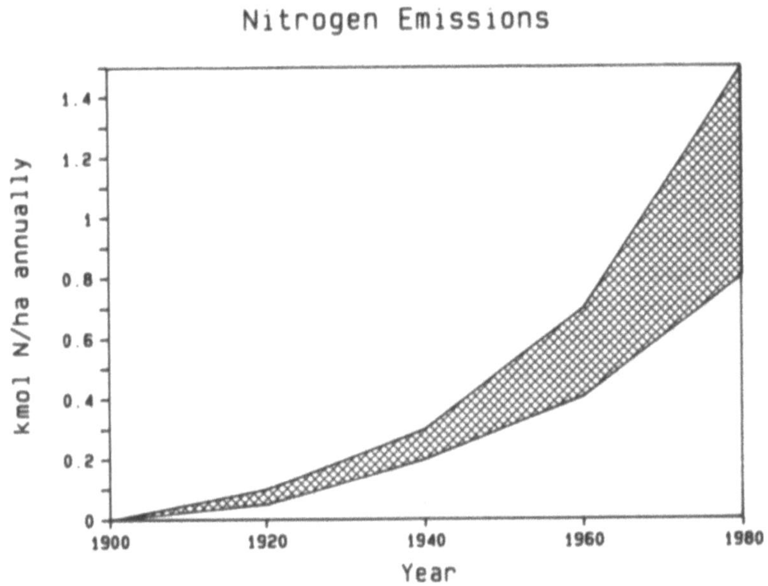

Figure 3.2. Nitrogen emission estimates for the South; the shaded area represents the degree of uncertainty. (After Husar 1986, reprinted by permission of the National Academy Press.)

of nitrogen oxides were very low (fig. 3.2). By 1940, sulfur emissions had climbed to about 0.30 kmol/ha, and nitrogen oxide emissions reached about 0.20 kmol/ha. Assuming these ratios reflected the composition of rainfall acidity, sulfuric acid (with 2 H^+/mol) would have contributed about 75% of the acidity. In 1980, sulfur emissions had climbed to about 1.2 kmol/ha, roughly matched by nitrogen emissions; about two-thirds of the acidity in precipitation was due to sulfuric acid. These estimates have a fair degree of uncertainty, but they document the rapid increase of N emissions through this century, including the recent trend toward continued increases. In contrast, S emissions have leveled off or even decreased somewhat in the country as a whole (NAPAP 1985).

Modes of Deposition

Rainfall contains chemicals that are dissolved in the water as well as small particles that are suspended in the water. Chemicals can also be deposited onto forest canopies with dew or in fog (if the wind is blowing). During dry periods, small particles (aerosols) settle out of the atmosphere. Sulfur dioxide gas can be deposited onto wet leaf surfaces or even assimilated through leaf stomata. Some nitric acid vapor may also be present in the air and be deposited directly onto canopies.

This diverse array of deposition pathways makes it almost impossible to

Table 3.2 Atmospheric deposition rates near Oak Ridge National Laboratory, Tennessee ($kmol_c$/ha annually)

Process	Sulfate	Nitrate	H^+	Ammonium	Calcium	Potassium
Precipitation	0.70	0.20	0.69	0.12	0.12	0.09
Dry deposition						
Particles	0.26	0.08	0.03	0.04	0.31	0.01
Vapors	0.62	0.26	0.85	0.01	0.00	0.00
Sum	1.58	0.52	1.57	0.17	0.43	0.02

Lindberg et al. 1986, Reprinted by permission of AAAS,

obtain precise estimates of total atmospheric deposition rates. Most collections simply sample the wet deposition that is obtained with rainfall captured in buckets or funnel collectors. If these collectors are also open to receive aerosol deposition during dry periods, the collection is referred to as bulk deposition. Dry deposition is easy to measure in a bucket or funnel collector, but the amount collected by these surfaces may not represent the amount trapped by the high surface area of forest canopies. One approach to estimating dry deposition is to measure the canopy surface area, the concentration of the particles (or gases), and then estimate a rate of transfer (called a deposition velocity). The deposition velocity is a largely empirical number that can vary greatly among forests and even through time within forests.

The most thorough characterization of deposition to a forest is probably the one published recently by scientists from Oak Ridge National Laboratory for a chestnut oak/white oak forest in Tennessee (Lindberg et al. 1986). The deposition of strong acid anions in wet precipitation was about 77% sulfate and 22% nitrate (table 3.2). However, only about 75% of the charge of these anions was balanced by H^+, the rest was associated with ammonium, calcium, and potassium. The rate of dry deposition was greater than wet deposition for all ions except ammonium. About 56% of the sulfate and H^+ came in dry, as did about 63% of the nitrate.

The Tennessee Valley region around Oak Ridge is probably more polluted than the South as a whole, but few intensive studies have been completed in the region. The ratio of dry deposition to wet deposition can vary substantially, both with location and season (fig. 3.3). The dry deposition rates across the South as a whole are probably lower than the Oak Ridge values. However, high-elevation sites, such as the Smoky Mountains, receive large inputs of fog precipitation and may have greater deposition rates than the Oak Ridge area.

Atmospheric Deposition in the South

Monitoring networks for precipitation chemistry were established in North America in late 1970s, with several systems in the United States and Canada. These systems have been coordinated into one large data base (the Acid Deposi-

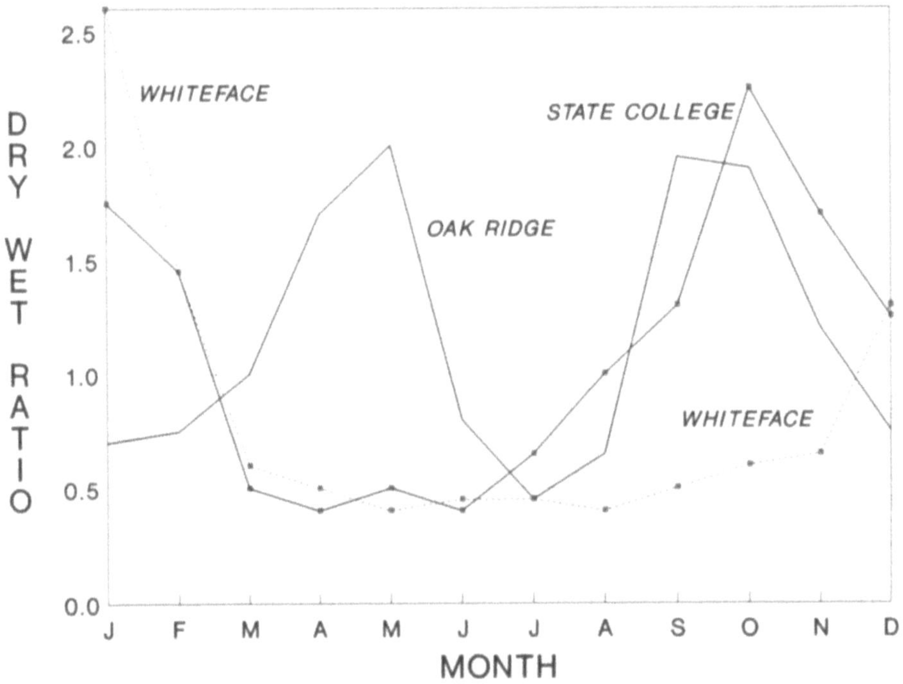

Figure 3.3. Seasonal dynamics in the ratio of dry to wet deposition for (1) Oak Ridge, Tennessee; (2) State College, Pennsylvania, and (3) Whiteface Mountain, New York (From NAPAP 1986.)

tion System) managed by Battelle's Pacific Northwest Laboratory in Richland, Washington. Most of the information from the various networks includes only wet precipitation or bulk precipitation, but monitoring of dry deposition is expanding.

Wet precipitation averages about pH 4.4 to 4.6 across the South, which translates into a deposition rate of about 0.2 to 0.4 kmol H^+/ha annually (fig. 3.4; NADP/NTN 1987). The pH in the northeastern United States is about 0.2 units lower, and the deposition rate about 50% greater. Sulfate deposition in the southeast is about 0.3 $kmol_c$/ha, compared with about 0.5 $kmol_c$/ha for the northeastern United States (fig. 3.5) and up to 3 $kmol_c$/ha in the most polluted parts of Europe. Nitrate deposition is about 0.15 $kmol_c$/ha in the southeast (fig. 3.5), compared with about 0.3 $kmol_c$/ha in the northeast. Ammonium deposition is about 0.10 $kmol_c$/ha in the sourtheast (fig. 3.6), and about 50% greater in the northeast. The deposition rate of calcium is similar along the eastern United States, averaging about 0.10 $kmol_c$/ha annually (fig. 3.6).

An alternate approach to estimating maximum rates of deposition is based on emission rates. The chemicals released from each source disperse across the landscape; the distance traveled before deposition varies with the chemical and weather conditions. Based on the emission figures discussed earlier, a maximum

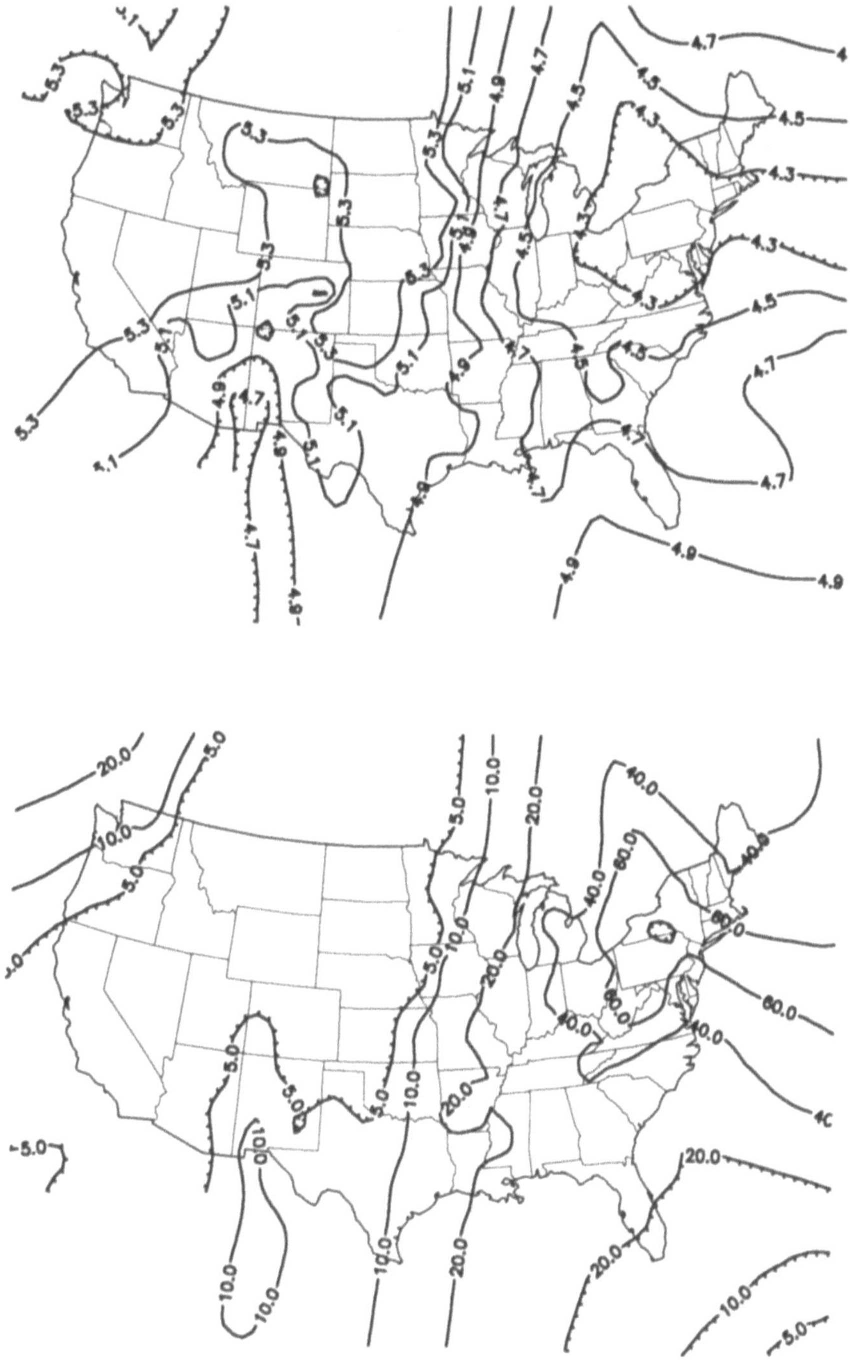

Figure 3.4. (*top*) Isolines of precipitation pH and (*bottom*) H^+ deposition (in g/m^2) for 1986. (From NADP/NTN 1987.)

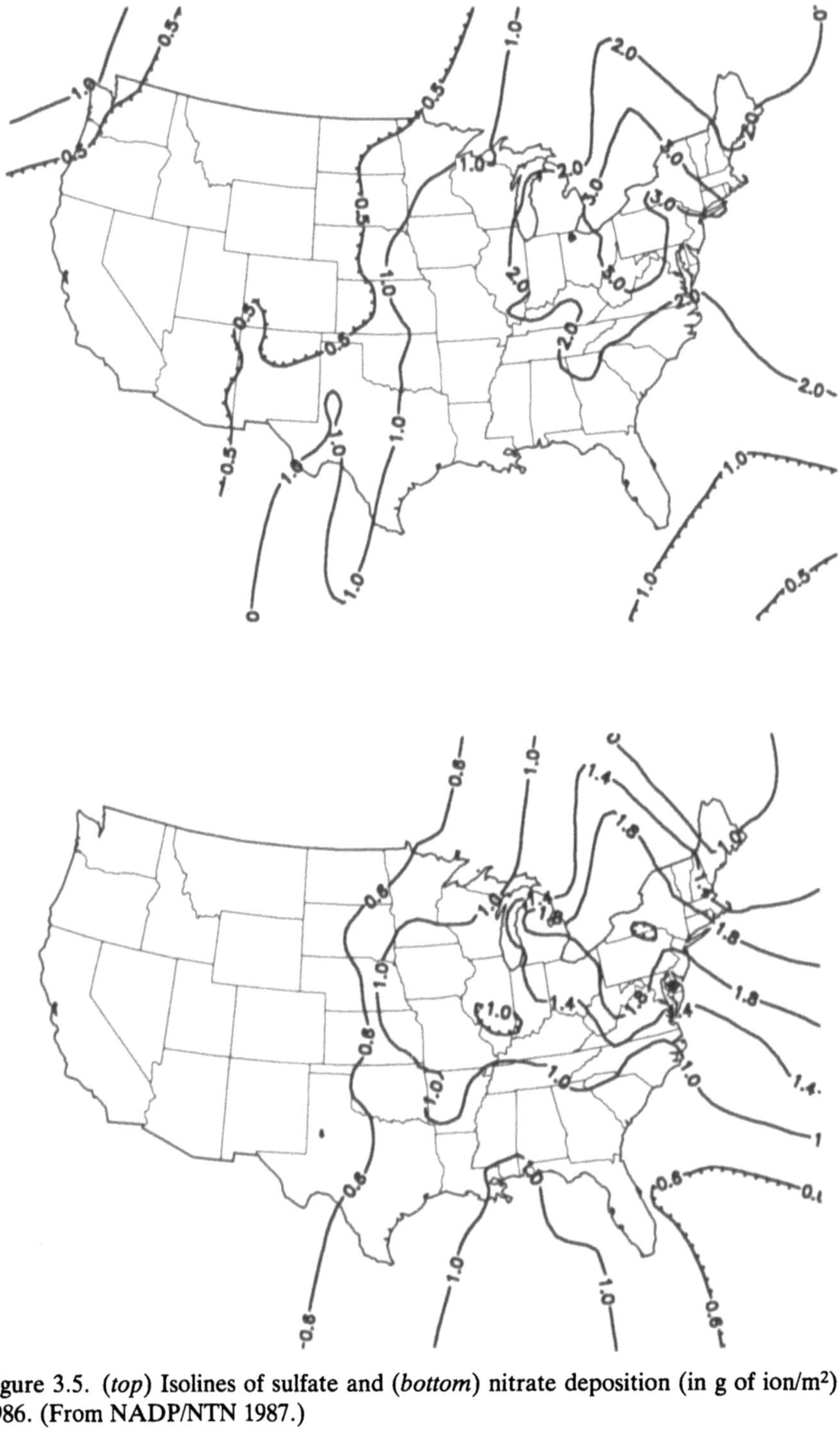

Figure 3.5. (*top*) Isolines of sulfate and (*bottom*) nitrate deposition (in g of ion/m²) for 1986. (From NADP/NTN 1987.)

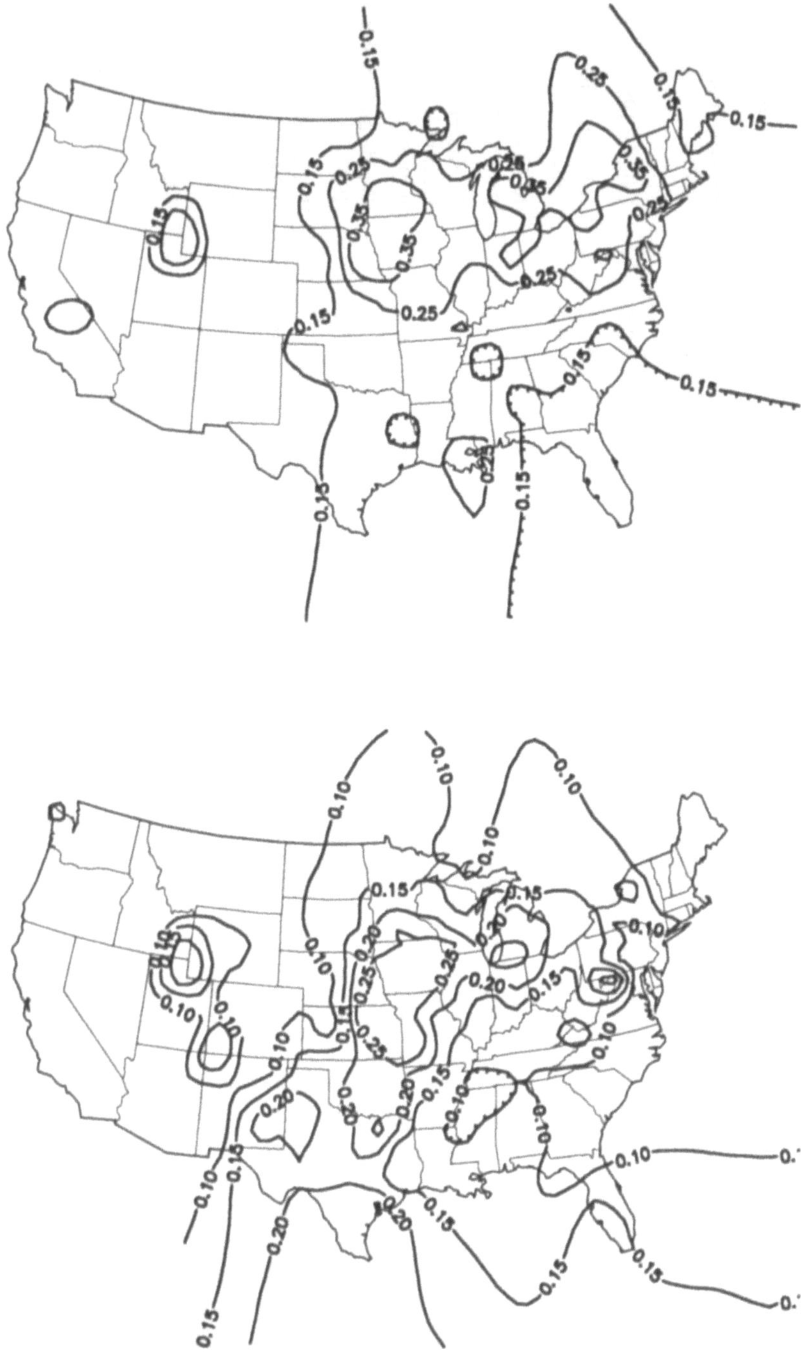

Figure 3.6. (*top*) Isolines of ammonium and (*bottom*) calcium deposition (in g of ion/m^2) for 1986. (From NADP/NTN 1987.)

Table 3.3. Input ($kmol_c$/ha annually) estimates from ecosystem nutrient budgets

Location/ Vegetation	SO_4^{2-}	NO_3^-	NH_4^+	Ca^{2+}	Mg^{2+}	K^+	H^+	Reference
Bradford Forest, FL, slash pine	—	0.45	0.26	0.76	—	0.15	—	Riekirk et al. 1978
Santee Watershed, SC, loblolly pine	0.50	→0.12←		0.26	0.13	0.03	0.5	Gilliam 1983
Clemson, SC, loblolly pine	0.77	0.06	0.22	0.14	0.08	0.06	0.5	Johnson et al. 1985a
Sandhills, NC, loblolly pine	1.12	0.77	—	—	—	—	0.4	K. Crawford, pers. comm. 1987
Duke Forest, NC loblolly pine	0.67	0.17	0.07	0.10	0.04	0.30	0.2	K. Knoerr, D. Binkley, unpub. data
Coweeta, NC, mixed hardwoods	0.67	0.22	0.15	0.23	0.08	0.05	0.6	Waide and Swank 1987
Chesapeake Bay, MD, mixed hardwoods	0.95	0.41	0.24	0.15	0.11	0.06	0.9	Weller et al. 1986
Catoctin Mountains, MD, mixed hardwoods	0.72	0.32	—	0.16	0.05	0.01	1.1	Katz et al. 1985
Soldier's Delight, MD, mixed hardwoods	0.90	0.37	0.31	0.25	0.08	0.04	—	Cleaves et al. 1974
Pond Branch, MD, mixed hardwoods	—	0.37	—	—	0.16	0.08	0.04	Cleaves et al. 1974
Shenandoah Park, VA, mixed hardwoods	0.56	0.23	0.13	0.12	0.04	0.02	0.5	P. Ryan, pers. comm. 1987

Fernow Forest, WV, mixed hardwoods	0.83	0.43	0.55	0.40	—	—	0.8	Helvey and Hunkle 1986
Walker Branch, TN, chestnut oak	0.92	0.14	0.09	0.35	0.08	0.05	0.7	
Loblolly pine	0.82	0.09	0.05	0.32	0.06	0.03	0.7	Johnson et al. 1985a
Camp Branch, TN, mixed hardwoods	1.25	0.31	0.42	0.49	0.11	0.07	0.2	Johnson et al. 1985a

estimate of total sulfur deposition would be 2.4 $kmol_c$ sulfate/ha and 1.2 $kmol_c$ nitrate/ha. These emission rates are even higher than the deposition estimates for the Oak Ridge region, and probably indicate that at least about half of the emitted S and N are indeed exported to the northeast and to the Atlantic Ocean. For the eastern United States as a whole, current estimates are that about 30% of the emitted sulfur dioxide is exported to the Atlantic Ocean (NAPAP 1986); the rest is either exported to Canada or deposited within the United States.

We also have summarized input rates published in various nutrient cycling studies in the South (table 3.3). The measurement techniques (and perhaps quality control) varied among these studies, so precice comparisons among studies are not warranted. Some broad conclusions are possible. First, sulfate deposition rates (bulk precipitation for most studies) fell between about 0.50 to 1.25 $kmol_c$/ha annually, which is notably greater than the 0.3 $kmol_c$/ha regional average from the NADP network. Nitrate ranged from 0.06 to 0.43 $kmol_c$/ha annually, compared with the regional NADP average of 0.15 $kmol_c$/ha. Ammonium was more variable, ranging from about 0.05 to 0.55 $kmol_c$/ha annually, in contrast with the NADP average of only 0.10 $kmol_c$/ha. All of these deposition estimates are subject to errors associated with sampling methodology, collection frequency, and analytical accuracy, thus clear interpretations of differences among data sources are not possible.

It is not evident how important the contribution of dry deposition is to the total acid input in the region. Based on the Oak Ridge study, dry deposition might be roughly equivalent to wet deposition. Some preliminary data from several sites of the Integrated Forest Study (funded by the Electric Power Research Institute through Oak Ridge National Laboratory) do indicate that dry and wet deposition may be roughly equal across the region (K. Knoerr, pers. comm. 1987). Current research will improve estimates of the importance of dry deposition at other locations in the region.

Trends in Deposition Rates

Annual variations in deposition rates make it difficult to assess long-term trends, and no clear inferences are possible from current information. Based on precipitation network data for seven locations in the South from 1978 through 1985, sulfate deposition decreased about 35% at four locations, with no change evident at the other three (NAPAP 1986). The trend for nitrate was less consistent; three sites showed no trend, two increased by about 30%, and two decreased by about 25%.

A ten-yr record is available for the U.S. Forest Service's (USFS's) Coweeta Hydrologic Laboratory in the mountains of western North Carolina (Waide and Swank 1987, Swank and Waide 1988). The average annual sulfate concentration has varied by up to twofold, with no long-term trend. Nitrate concentrations were less variable and showed a hint of an increasing trend that was not statistically significant. Despite the absence of a trend in sulfate input, there was a significant increase in sulfate concentration (fig. 3.7) and export from two

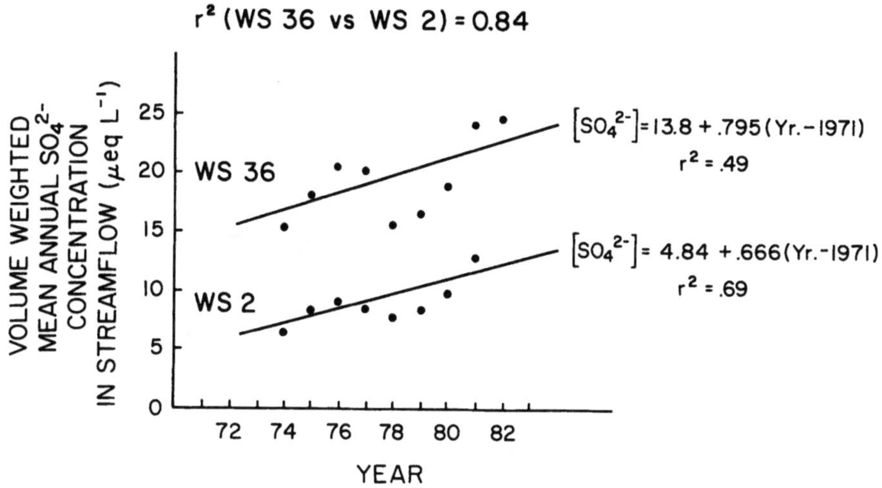

Figure 3.7. The concentrations of sulfate in streamwater at Coweeta have increased by about 0.7 μmol/L annually between 1972 and 1982. (From Swank and Waide, 1988.)

Coweeta watersheds. The vegetation and treatment histories of these watersheds differ, so the similar trend across watersheds suggests that cumulative inputs might underlie the trend; however, the data cannot clearly support precise arguments in any direction.

The computer simulations in chapter 7 showed that acidification trends in representative forest soils in the South were fairly sensitive to deposition rates. Therefore, assessments of soil acidification in the South will become more reliable as better deposition information becomes available.

4. Forest Soils of the South

The impacts of acidic deposition depend strongly on the specific chemistry of each soil and the nutrient cycles within each ecosystem. To provide perspective on the general characteristics, patterns, and diversity of forest soils in the South, the general geomorphology of the region and the major features of the most important and extensive soil groups are described herein.

Two spatial scales are important in characterizing the geomorphologic setting of soils. The regional or physiographic perspective accounts for broad climatic and geologic features that shape soil formation, including temperature, precipitation, and substrate geology. This chapter focuses primarily on these broad patterns. The second scale of soil geomorphology accounts for local topographic patterns. Local geomorphology, hydrology, and geology affect the inputs, outputs, and internal transfers of a soil pedon. The angle and length of slopes influence the hydrology of pedons, as does the distance between drainageways. For example, broad, nearly level areas can pond water even if the soils have adequate infiltration and permeability properties (such as broad, low-elevation interfluves and broad river terraces). Thorough understanding of specific sites is not possible without the context of the local geomorphology in which it occurs.

Climate

The climatic setting of the southeastern United States is diverse, but it is generally characterized as having a continental climate merging into a maritime climate along the Coastal Plain. Annual precipitation ranges from about 1000 to 1625 mm (fig. 4.1). Precipitation is markedly lower near the western margin, where woodland vegetation grades into grasslands. Greatest precipitation occurs in several coastal areas and in the mountainous regions. Mean annual air temperature (fig. 4.2) and potential evapotranspiration (fig. 4.3) are markedly higher in the southern areas and at low elevations. The interaction of precipitation and evapotranspiration produce a range of surface runoff rates across the region (fig. 4.4).

Regional Geomorphology and Soils

The southeastern United States (Alabama, Arkansas, Florida, Georgia, Louisiana, Mississipi, North Carolina, Oklahoma, South Carolina, Texas, and Virginia) can be grouped into ten regions (Fenneman 1983, Thornbury 1965), each encompassing a range of climatic, geologic, and soil characteristics. An overview of these ten regions is followed by a discussion of the implications for soil acidification.

Coastal Plain

The Coastal Plain is the largest region, extending over a broad region from Texas to Virginia (fig. 4.5). The region is generally characterized by low relief, with broad flats and level uplands at low elevations. Several subdivisions of the Coastal Plain are important.

The Lower Coastal Plain (lowlands or flatwoods) consists of virtually level lowlands interspersed with swamps, estuaries, and lagoons. These lowlands are poorly dissected, have minimal relief, and elevations of less than about 30 m. They typically have long, nearly level slopes, with poor soil drainage at the center of the broad flats. Soil materials are dominated by river and marine sediments of sand, silt, and clay (Murray 1961). Soils are largely Ultisols (Hapludults, Ochraqults, Paleudults) (USDA—Forest Service 1969, Soil Survey Staff 1975, Buol et al. 1980, Buol 1983). Near-shore areas along the Atlantic Coast have extensive beach and dune sediments where Entisols are dominant (Quartzipsamments, Psamments, Sulfaquents) (Daniels et al. 1984). The Lower Coastal Plain is about 90% woodland (USDA—Forest Service, 1969).

The Upper (or Interior) Coastal Plain is characterized by broad uplands or low-lying plateaus. This area is geologically older than the Lower Coastal Plain, and it exhibits greater dissection, relief, and elevations (30 to 200 m). Soils are primarilty Ultisols (Paleudults, Paleaquults).

Across the South, the Coastal Plain region is divided into two areas by the Mississippi River, and each of these areas can, in turn, be subdivided. The Eastern Coastal Plain extends from Virginia through Mississippi. It is bounded by the Atlantic Ocean, the Gulf of Mexico, and the Mississippi River Valley system. The interior margin is clearly defined topographically from Virginia to Georgia by a sequence of escarpments or sandy hills of higher elevation that diminish to

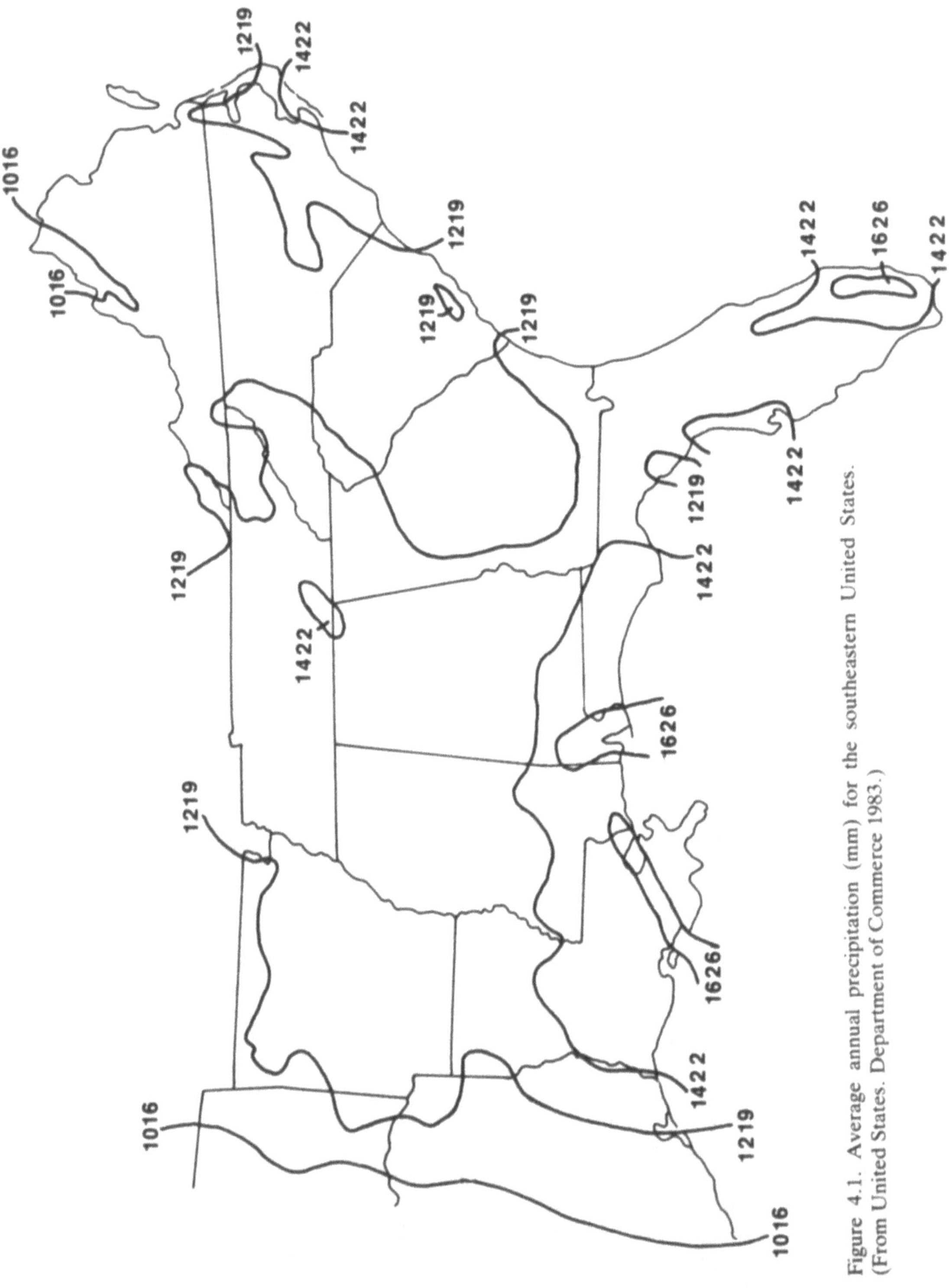

Figure 4.1. Average annual precipitation (mm) for the southeastern United States. (From United States. Department of Commerce 1983.)

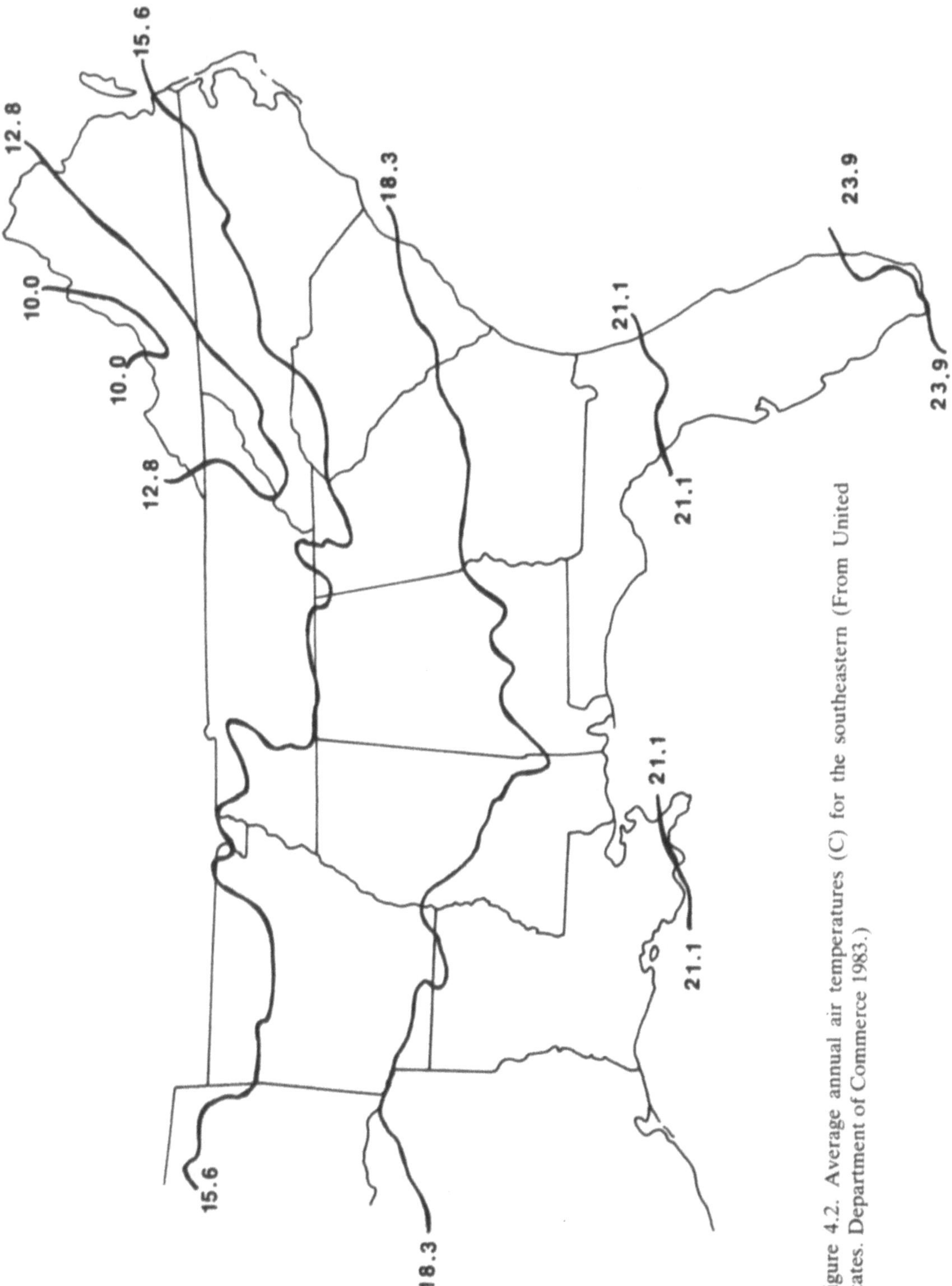

Figure 4.2. Average annual air temperatures (C) for the southeastern (From United States. Department of Commerce 1983.)

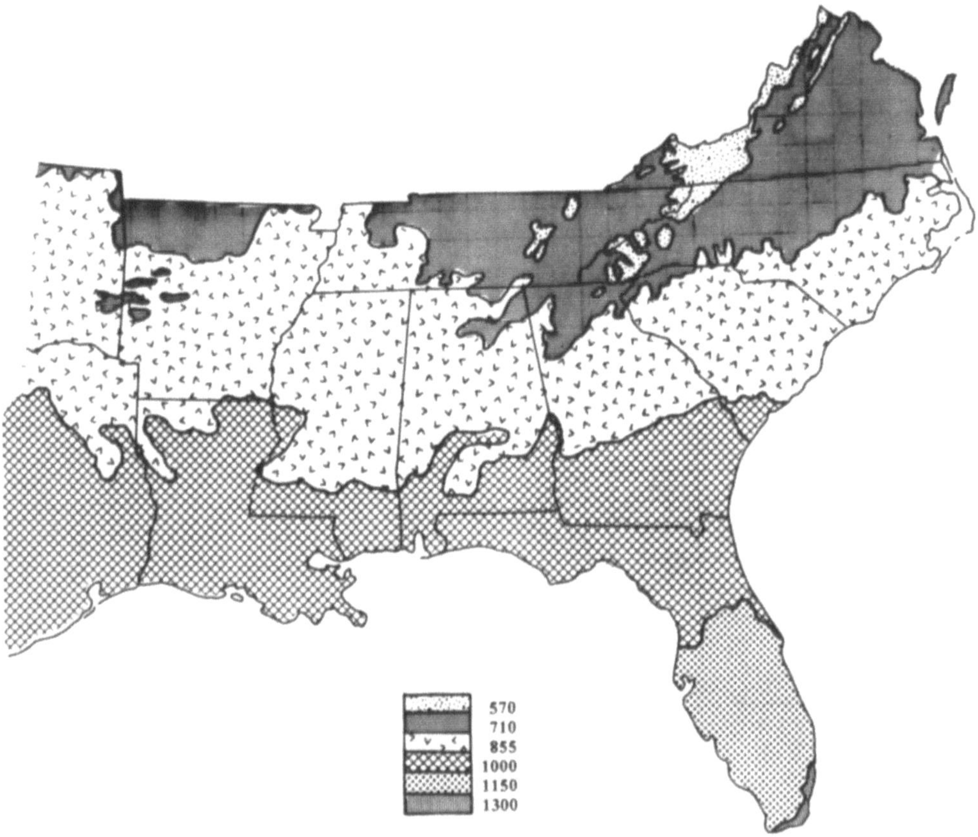

Figure 4.3. Average annual potential evaporation (mm) for the southeastern (From United States. USDA—Forest Service 1969.)

the west. The sand hills are locally extensive and dominated by Entisols (Quartzipsamments) interspersed with Ultisols (Paleudults, Fragiudults). The sand hills are about 60% woodland. The Tertiary and Cretaceous strata that form the narrow, definintive escarpments of the Upper Coastal Plain in the east, broaden westward as the Tertiary sediments narrow (Murray 1961).

The Coastal Plain of west Florida, Alabama, and Mississippi is sometimes referred to as the Eastern Gulf Plain (Buol 1983) and also exhibits a westward broadening in a belted complex of low-lying cuestas with gentle slopes facing seaward and short, steep slopes facing the interior. This complex is dominated by finer textured Ultisols (Paleudults, Hapludults, Fragiudults, Paleaquults). The greater geologic diversity is evident through the occurrence within the belts of associated Alfisols (Paleudalfs) derived from limestone, and in northeastern Mississippi Vertisols (Chromuderts) derived from shales. This portion of the Upper Coastal Plain is about 50% to 75% woodland.

The western edge of the Coastal Plain along the Mississippi River system is

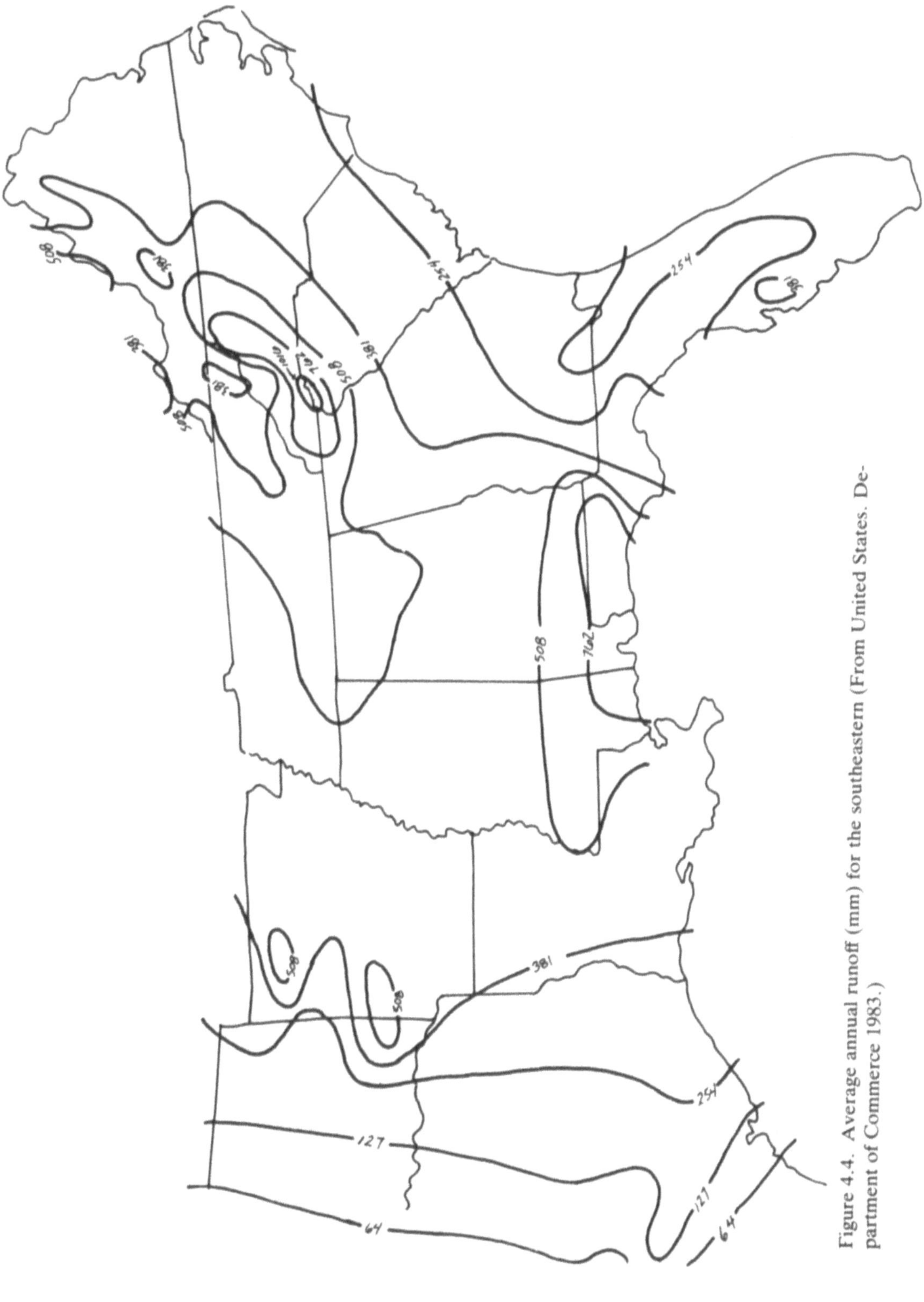

Figure 4.4. Average annual runoff (mm) for the southeastern (From United States. Department of Commerce 1983.)

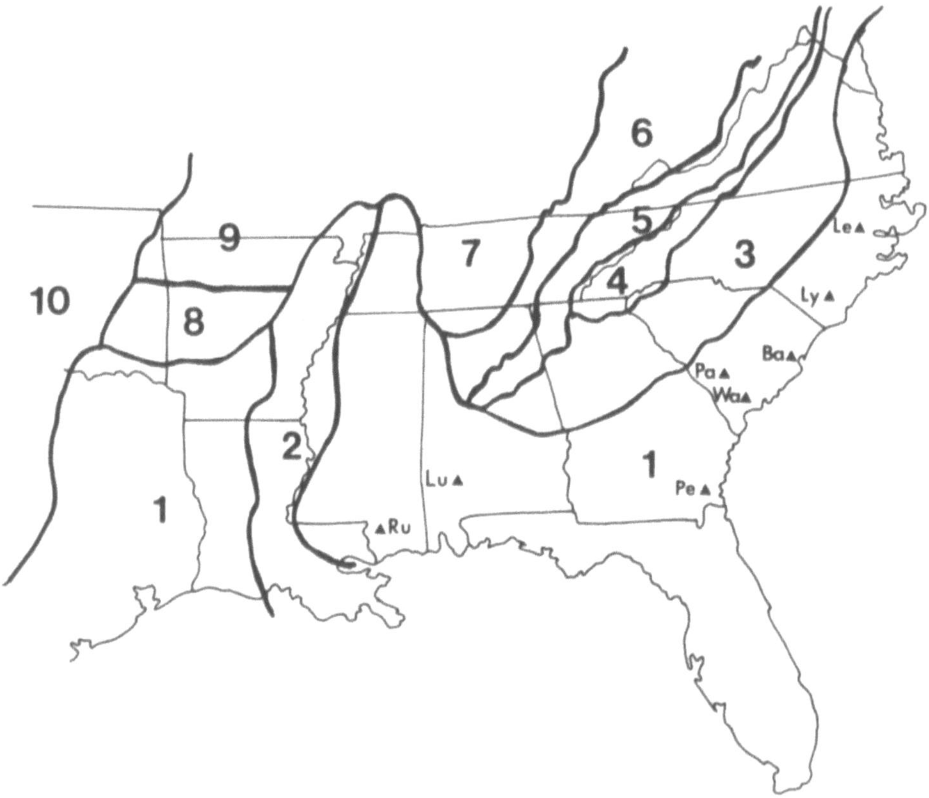

Figure 4.5. Geomorphic regions of the southeastern United States (after Murray 1961, Thornbury 1965): (1) Coastal Plain, (2) Mississippi River Valley, (3) Piedmont, (4) Blue Ridge, (5) Ridge and Valley, (6) Appalachian Plateau, (7) Interior Low Plateau, (8) Ouachita Uplands, (9) Ozark Plateau, and (10) Central Lowlands. Triangle mark common locations of the soil series (abbreviated with first two letters) used in the simulations in chapter 7.

mantled on the surface by up to 20 m of windblown silt that becomes progressively shallow toward the east. The resulting soils are largely Alfisols (Hapludalfs, Fragiudalfs) interspersed with Ultisols (Paleudults). This loess-covered western edge of the Upper Coastal Plain is about 25% woodland.

The Florida penninsula is a distinct subregion. The center of the peninsula is dominated by sandy sediments overlying a broad, gently rolling limestone upland, with elevations of 20 to 80 m and minimal relief. Soils are dominantly Entisols (Quartzipsamments, Psamments, Haplaquents), Spodosols (Haplaquods), and Alfisols (Hapludalfs) where limestone occurs near the surface. About 40% of the interior of Florida is woodland. The eastern one-third of the peninsula is a continuation of the low-elevation, low-relief flatwoods of the Atlantic Coastal Plain. The deep, sandy marine sediments give rise to Spodosols

(Haplaquods) interspersed with Entisols (Psamments). Organic soils (Medisaprists, Medihemists) are locally extensive, especially at very low elevations south of Lake Okeechobee. These composite lowlands are about 25% woodland.

The western Gulf Coastal Plain lies west of the Mississippi River Valley and is similar in geomorphology to the eastern Gulf Coastal Plain. The Lower Coastal Plain is a low-elevation, low-relief plain covered by fine alluvium and marine sediments near the Mississippi River Valley, giving way to sandier Pleistocene sediments to the west. The Upper Coastal Plain is again a belted complex of low-lying cuestas of Tertiary and Cretaceous sedimentary strata. Further west are Permian and Pennsylvanian sedimentary rocks.

Despite similar geomorphology and elevations to the eastern Gulf Plain, major soil groups of the western Gulf Plain occur in two distinct zones. Areas of Louisiana and eastern Texas are similar to the eastern Gulf Plain. Soils are dominately Ultisols (Paleudults, Hapludults) and Alfisols (Paleudalfs). About 60% of the area is woodland. Further west, soils deviate from the regional pattern owing largely to a drier climate and changes in geology. The Lower Coastal Plain is dominated by Mollisols (Haplaquolls) and Vertisols (Pelluderts). The more rolling Upper Coastal Plain of the Texas claypan area is dominated by Alfisols (Ochraqualfs, Paleudalfs, Hapludalfs, Albaqualfs). About 50% of the area is woodland. Vertisols (Pellusterts, Chromusterts) dominate the Texas blacklands, very little woodland is present.

Mississippi River Valley

The Mississippi River Valley region splits the southeastern United States. It is a low-elevation (0 to 170 m), low relief (nearly level) complex of river terraces, meanders, levees, and backwater swamps. It is a composite of alluvium from the Mississippi River and its tributaries, overlying older marine sediments. A thick mantle of windblown silts occur on higher-elevation terraces.

The active floodplain is composed largely of Entisols (Fluvaquents in the active floodplains, Hydraquents in quiescent areas). The lower alluvial terraces are largely Inceptisols (Haplaquepts). Mid-elevation terraces are dominantly Alfisols (Ochraqualfs) with intermingled Mollisols (Haplaquolls). The older, highest-elevation terraces in northeastern Louisiana and eastern Arkansas are thickly covered by loess and are dominantly Alfisols (Fragiudalfs, Glossaqualfs, Albaqualfs). All of these soils should be resistant to acidification.

Piedmont

The Piedmont region is the second-largest physiographic unit in the southeastern United States. This rolling erosional plain consists of moderate elevation (100 to 350 m), minor relief (10 to 30 m) uplands derived from diverse metamorphic and igneous rocks. It parallels the Appalachian Highlands from Georgia through

Virginia, sloping gently to the Coastal Plain on the east and south. The old erosional nature of this region has resulted in soils dominated by: Ultisols (Hapludults, Paleudults, Rhodudults, Paledults) on acidic rocks; some Alfisols (Hapludalfs, Paleudalfs) on basic igneous rocks; and some Inceptisols (Dystrochrepts) on steep slopes. Within this region are several locally extensive areas derived from Triassic age, metavolcanic and marine sediments. These Triassic basins are dominated by Alfisols (Hapludalfs). About 60% of the region is woodland.

Blue Ridge

The Blue Ridge mountain range, with a wide range in elevation (300 to 1200 m with peaks exceeding 2000 m) and considerable relief, is composed of predominantly igneous and metamorphic rocks. Moderately steep to steep slopes are extensive and dominated by Inceptisols (Dystrochrepts). Less steep areas with more stable slopes in valleys or broader ridgetops are dominated by Inceptisols (Haplumbrepts) and Ultisols (Hapludults). About 60% of the region is woodland.

Ridge and Valley

A series of narrow geologic belts of moderate elevation (200 to 1000 m) and significant relief continues to parallel the northeast–southwest regional trend of the Appalachians. It slopes gently to the south and west. The alternating belts vary in width and are comprised of valleys bounded by steep ridges. The rolling-to-hilly limestone-and-shale valleys are bounded by steep mountain ridges of more resistent sedimentary rocks. Soils are dominantly Ultisols (Paleudults, Hapludults) with some Alfisols (Paleudalfs) over limestone areas. The less stable slopes and steep ridges are dominantly Inceptisols (Dystrochrepts). About 40% of the region is woodland.

Appalachian Plateaus

The interior plateaus of the western Appalachian Uplands are gently rolling to rolling, deeply dissected, and at moderate elevation. The region is bounded by steep escarpments (300 to 670 m) falling away to the west and south, except at the mountainous eastern edge which exceeds 1200 m. The plateaus are composed of sedimentary rocks of nearly level sandstone-capping strata and eroded shale strata. The more stable, nearly level-to-rolling uplands and valley floors are dominated by Ultisols (Hapludults, Paleudults). The steep escarpments and valley walls are predominantly Inceptisols (Dystrochrepts). About 80% of the region is woodland.

Interior Low Plateau

The rolling, low-lying interior plateau consists of a broad, extensively dissected rim surrounding a less dissected central core. It is lower in elevation and west of the Appalachian plateaus. Elevations are moderate (100 to 300 m) with significant relief on steep slopes occurring in the dissected rim and the more subdued relief in the rolling central plateau. The geology consists of diverse sedimentary rocks but is strongly influenced by limestone. Alfisols (Hapludalfs, Paleudalfs) dominate the less dissected plateau interior with Ultisols (Paleudults, Fragiudults, Hapludults) interspersed with Alfisols (Hapludalfs) dominating the dissected rim. About 25% of the region is woodland.

Ouachita Uplands

The Ouachita Uplands is a compact physiographic region that lies west of the Mississippi River and entails two subregions: the Ouachita Mountains and the Arkansas River Valley. The Ouachita Mountains dominate this region and are similar in form to the western Appalachian Uplands (Fenneman 1938). Here the alternating ridge and valley structures parallel the east–west trend of the Ozark Mountain complex to the north, to which they are structurally related. Elevations are moderate (100 to 600 m) with some peaks exceeding 700 m. The relief is significant (100 to 300 m) on the steep-sloped ridges and valley walls. The area gradually falls in elevation southward to the Coastal Plain. The sedimentary rocks exhibit steep sandstone or cherty-limestone ridges and rolling hills dipping to the south, dominated by Ultisols (Hapludults). The steeper north-facing valley slopes and escarpments are primarily Inceptisols (Dystrochrepts). The rolling valley areas of weathered slate, shales, and sandstones are predominantly Ultisols (Hapludults). About 80% of this subregion is woodland.

The Arkansas River Valley system and its associated ridges are elevationally similar to the Ouachita Mountains, and it separates this region from the Ozark Mountains to the north. Valleys and ridges are broader with similar soils except for the occurrence of Alfisols (Paleudalfs) that appear on the more stable positions where limestone is near the surface. The Arkansas River Valley itself is a gently sloping-to-rolling dissected plain dominated by Alfisols (Hapludalfs) intermingled with Ultisols (Fragiudults) and Inceptisols (Haplaquepts in low-lying areas, Dystrochrepts on steep slopes). About 40% of this region is woodland.

Ozark Plateau

The high, deeply dissected Ozark plateaus of sedimentary rocks are characterized by narrow rolling ridgetops falling away to steep sideslopes of considerable relief. The northern area has moderate elevations (170 to 500 m) with significant relief (100 to 200 m). The dissected sandstone ridgetops and rolling uplands are predominantly Ultisols (Paleudults, Fragiudults). Alfisols (Paleudalfs) dominate the valley floors overlying limestone, dolomite, and shales. Inceptisols (Dystrochrepts) dominate steep side slopes and valley walls. About 60% of this area is woodland. The southern area has similar geomorphology but is more deeply dissected with greater elevations (170 to 800 m) and relief (up to 300 m). The sandstone ridges and valley bottoms are primarily Ultisols (Hapludults, Paleudults) with Inceptisols (Dystrochrepts) on steep slopes. About 75% of this area is woodland.

Central Lowlands

The Central Lowlands area of low elevation and low relief is characterized by broad flats, cuestas, and rolling hills. It lies between the treeless plains to west and the Ozark/Ouachita Uplands and the Coastal Plain to the east. There are several subregions.

The eastern Cherokee Prairies have low elevations (130 to 400 m) and low relief (less than 50 m) on Pennsylvanian-aged sedimentary rocks. The sandstones, shales, and clays are dominated by Mollisols (Paludolls, Argiudolls) and interspersed with Alfisols (Albiqualfs, Hapludalfs) in alluvial areas. About 10% of this area is woodland.

The western Cross Timbers area has similar elevations (200 to 400 m) but is more deeply dissected, with subsequently more relief. The largely sandstone and limestone strata are dominated by Mollisols (Paleustolls, Argiustolls, Haplustolls) and Alfisols (Paleustalfs, Haplustalfs). This area is about 25% woodland.

The remaining physiographic regions of the southeastern United States lie to the west of those described, do not have significant woodlands, and are not dealt with here.

Importance to Soil Acidification

We would like to be able to tabulate the forested land areas in each of the major soil groups in the South. However, the current state of the multiple data bases that contain such information does not allow simple merging to produce such information. The recent work by Turner et al. (1986) remains the state of the art. However, we can make some inferences about the major features controlling soil acidification and how these relate to soil taxonomy.

From the background in chapter 2 and our computer simulations in chapter 7, three factors probably account for the majority of differences among soils in acidification potential: sulfate retention through adsorption reactions, exchangeable basic cations, and mineral weathering.

We think it is premature to try and relate sulfate adsorption to soil taxonomic classes. Earlier work by D.W. Johnson et al. (1980) and Olson et al. (1982) attempted to relate sulfate adsorption capacity to soil orders. However, more recent work has shown that sulfate adsorption may not follow the expected pattern across soil orders. D.W. Johnson et al. (1986) found that sulfate was not retained in some Ultisols in Tennessee but was retained by Inceptisols in

Washington. It would be premature to conclude that all Ultisols in the South show poor sulfate retention; given that Inceptisols is one of the most variable soil orders (Soil Survey Staff 1975), generalization about sulfate retention by Inceptisols may not be warranted. Our computer simulations showed (along with earlier arguments and studies) that geochemical sulfate adsorption would retard soil acidification from acidic deposition only until a new steady state with respect to higher sulfate concentrations was attained. Therefore, we conclude: (1) sulfate adsorption may be broadly related to soil order (or other taxonomic group), but the relationship is not yet clear; and (2) sulfate adsorption may account for short-term differences in rates of soil acidification but will probably not be important beyond this initial difference in delay periods.

Most soil acidification discussions have considered the pools of exchangeable basic cations and the rates of mineral weathering to be the major factors buffering inputs of strong acids. In most of the 10 geomorphologic regions of the Southeast, Ultlisols and Alfisols were the dominant soil orders. The major distinctions between these orders are the base saturation of the exchange complex and the presence of weatherable minerals. Ultisols generally must have less than 35% base saturation (with CEC defined by sum of cations) and less than 10% of the 20 to 200 μm size fraction consisting of weatherable minerals. Alfisols generally have higher base saturation (>35%), pH, and more weatherable minerals. By the basic cation-leaching hypothesis of Turner et al. (1986), Alfisols may show higher rates of leaching of basic cations than Ultisols, but this characteristic would not be associated with drastic decreases in nutrient cation availability to plants or with increases in aluminum concentrations to toxic concentrations in soil solution. Although Entisols are by definition poorly developed soils, some (such as Psamments) may be low in base saturation and weatherable minerals, leaving them more poorly buffered than even Ultisols.

Our simulations (chap. 7) indicated that the long-term acidification of sensitive soils was determined largely by the balance between acid generation (or input) and mineral weathering. If H^+ input exceeded mineral weathering, the soils acidified, with the time course partially determined by other parameters. Given that Ultisols are defined in part by their low contents of weatherable minerals, they should be considered susceptible to acidification by acidic deposition unless other factors can be demonstrated to provide sufficient buffer capacity. For example, 98.5% of the minerals (to a depth of 80 cm) in the Ultisol at the Duke Forest site mentioned in chapter 2 were comprised of oxides of silica, aluminum, and iron. The consumption of H^+ and release of basic cations in such a soil should be very low. Interestingly, Ultisols with high contents of aluminum oxides should be very well buffered with respect to acidification near pH 4 because of weathering of these oxides. This buffering mechanism results in increases in concentrations of aluminum on the exchange complex and in drainage waters, however, and this may have deleterious effects on terrestrial and aquatic organisms.

Alfisols (such as the generic well-drained soils simulated in chap. 7) should be very resistant to acidic deposition, although they may exhibit increased losses of basic cations and changes in pH.

If Ultisols are very old, weathered, and relatively susceptible to acidification, why are they not more acidic than they currently are? Much of the answer lies in the changing deposition environment and changing land use and management practices. Ultisols are derived from diverse materials, but very few have many primary minerals that contain basic cations (other that some micas). Those few Ultisols that do have a supply of basic cations are intensively cultivated and are not in woodland. Most Ultisols were forested at the time of European settlement (Soil Survey Staff 1975); almost all forest soils in the South have been cleared and farmed for varying periods (from decades to centuries). Agricultural practices led to massive erosion in some areas—(an average of 17 cm across the South, Trimble 1974) and nutrient removal and depletion of organic matter in all areas. In the past 50 yr, fertilization of crops became common, and the fertility of many Ultisols was increased (Daniels 1987). Because of this varying land-use history, generalizations about Ultisols are limited. However, it is interesting to note that when the abandoned cotton field in South Carolina described in chapter 2 (see also Binkley et al. 1989) was planted to loblolly pine, soil pH declined by 0.3 to 0.8 units in just 20 yr. Over the same period, base saturation decreased from 40% to 7% in the A horizon; base saturation in the B horizon was depleted from 44% to 34% (Binkley et al. 1988). Agricultural practices increased soil pH and base saturation above the values defined for Ultisols, but less than three decades after agricultural management stopped, the soil had reverted to values more typical for Ultisols. No comparable data exist for other old fields, but the general declines in forest productivity are expected as the residual effects of agriculture wear off (Sheffield et al. 1985).

If the Ultisols had reached steady state with respect to the deposition environment in preindustrial times, they would still be expected to acidify further with increased deposition and increased ionic strength of soil solutions. Indeed, the geochemical retention of sulfate by adsorption results from this increase in soil-solution concentration; when steady state is attained with the higher concentrations of sulfate in solution, no net adsorption occurs. Soils that currently exhibit geochemical sulfate retention are probably not, by definition, in steady state with the current deposition environment.

The Duke Forest pine stand discussed in chapter 2 also presents a good example of residual effects of agricultural management. The net annual loss of sulfate from the soil came from a total pool (organic plus total mineral) of S of about 58 kmol/ha (1800 kg/ha), which is exceptionally high for very weathered Ultisols. This pool probably resulted from P fertilization with superphosphate [$3Ca(H_2PO_4)_2{\cdot}H_2O + 7CaSO_4{\cdot}2H_2O$], which contains 1.5 mols of S and 1.9 mols of Ca for every mol of P. Most currently forested soils in the South have experienced some degree of agricultural management, and the residual effects may overshadow expected trends in soil chemistry and response to acidic deposition.

We are confident that many forest soils in the South are probably not in steady state with respect to long-term soil development processes or with current levels of acidic deposition, and fairly substantial changes in soil chemistry should be expected. The magnitude, rates, and implications of these changes are not clear and should be addressed by specific research efforts (see chap. 8).

5. Previous Evaluations of Sensitivity of Southern Soils

Several studies have provided information on changes in soil chemistry for specific sites in the South, and others have evaluated the regional sensitivity of soils to strong acid inputs. In this chapter, we discuss the site-specific studies and then examine the previous investigations that have assessed the regional sensitivity of soils in the South to acidic deposition. We conclude by considering the criteria that have been used to define sensitive soils, and we develop our own definitions for use in the computer simulation runs in chapter 7.

Site-Specific Studies

The ideal investigation of soil acidification would be an examination of changes in soil chemistry over a long enough period of time for deposition impacts to appear. A variety of studies have reported decreases in pH of 0.5 to 1.0 units over periods of 2 to 10 decades (Stohr 1984, Brand et al. 1986, Johnston et al. 1986, Nilsson 1986, Tamm and Hallbacken 1986, Anderson 1987, Berden et al. 1987, Falkengren-Grerup 1987, Binkley et al. 1988). One of the best direct assessments comes from Sweden, where Tamm and Hallbacken (1986) resampled and analyzed soils in 1982 at the precise locations of 1927 samplings. They reported that pH in upper soil horizons had decreased by about 0.8 units in both beech and Norway spruce forests over the 55-yr period. Even deeper horizons acidified by about 0.5 units. The reason for this pH change is not evident, but some speculation may be useful. Recalling the factors that regulate soil pH (chap. 2), the acidification could have been due to a change in the size of the exchange complex. However, more than a doubling of the exchange capacity would have been required to produce this trend. The degree of dissociation of the exchange complex is a strong possibility—this shift could have been driven by either plant uptake or by inputs of strong acids from the atmosphere. If acidic deposition were the cause, however, the spruce stands might have been expected to show a greater change than the beech stands because of larger inputs of strong acids associated with coniferous vegetation (Ulrich, 1983). Alternatively, there may have been a greater accumulation of organic detritus under the spruce stand that buffered the inputs in strong acids. The strength of the acids that comprise the exchange complex could also have increased, but so little is known about such changes in forest ecosystems that is is difficult to estimate how likely this might be. The final possibility is an increase in the ionic strength of the soil solution (and precipitation) that undoubtedly occured, but this would probably not account for the entire change.

In the southern United States, there have been no 50-yr studies of changes in soil chemistry. Three studies fall within the 10- to 20-yr range: Carol Wells's pine plantation in South Carolina, the Walker Branch studies by Dale Johnson and colleagues at Oak Ridge National Laboratory, and the work of Wayne Swank, Jack Waide, and colleagues at the Coweeta Hydrologic Laboratory in western North Carolina.

Chapter 2 discussed a 20-yr stand development of a loblolly pine plantation in an abandoned agricultural field showed a change in pH of about 0.8 units in the upper soil and about 0.2 units decrease in the lower soil. These changes were associated with both decreased base and decreased acid strength of the complex. The importance of atmospheric deposition was probably smaller than within-ecosystem processes. All of the deposited nitric acid was probably assimilated (and neutralized) by vegetation or soil microbes. If deposition of sulfuric acid were near the regional average of 0.3 $kmol_c$/ha annually and none of it was retained, then atmospheric deposition might have accounted for about 15% of the acidification over 20 yr. This study demonstrated that changes in soil chemistry are to be expected, especially when the vegetation and management regime shift from agriculture to forestry and that the effects of acidic deposition need to be considered as additions to natural processes.

Intensive studies in the Walker Branch watershed at Oak Ridge National Laboratory have produced a record of soil chemistry from 1972 to 1982 (D.W. Johnson et al. 1987). During this period, the most notable changes in soil chemistry included decreases in exchangeable calcium and magnesium at 45- to 60-cm depths. Calcium declined by about 20 $mmol_c$/kg, and magnesium decreased by about 10 to 20 $mmol_c$/kg. They also examined seasonal dynamics in the pools of exchangeable cations and found that the change over ten years in the surface soil was not much greater than the variation found within a single year, but changes in the deeper horizon exceeded the slight annual variations. The researchers could not construct a complete nutrient budget for the ecosystem because soil samples were not collected in 1972 for the 15- to 45-cm depths.

However, the authors concluded that the reduction in calcium might be accounted for by accumulation in vegetation but that the magnesium was probably lost through leaching (soil-solution data were not available). These soils had low supplies of nutrient cations in 1972, and the reduction by 1982 raises serious concerns for the nutritional status of the forest. Implications for soil acidification, however, are not clear. Exchangeable aluminum did not change much at the 45- to 60-cm depth, as would be expected if the depletion of basic cations lead to increased soil acidity; high variability in the large pool of exchangeable aluminum may have obscurred real changes. The pH of some locations appeared to have increased slightly over the 10-yr period. Probably the most important finding of this study was the large variation in basic cation pools that occurred seasonally on soils with very low pool sizes. These pools also appeared to be reduced significantly by either vegetation uptake or leaching losses in just 10 yr. If this trend continued, changes in other pools (and in soil pH) might become measurable.

The Coweeta studies were mentioned in chapter 3 (fig. 3.7); no trend was apparent for sulfate deposition, but sulfate output in several watersheds clearly increased over the 10-yr record (Waide and Swank 1987). Interestingly, the output of bicarbonate decreased more (on a charge basis) than sulfate output increased, so the net export of cations decreased slightly. This trend has two implications. First, accelerated leaching of sulfate (which might be expected, as sulfate adsorption reaches a steady state with respect to atmospheric inputs) was not associated with increased losses of basic cations because of the reduction in bicarbonate export. Second, the decrease in bicarbonate probably resulted from a decrease in soil pH; chemical equilibrium calculations indicate the pH of soil solutions should have decreased by about 0.10 to 0.15 units. (Alternatively, large decreases in the CO_2 concentration of the soil could also explain the decrease in bicarbonate, but we would not expect the pCO_2 to show such a long-term trend.) A change of this magnitude in pH in 10 yr is important, but the longer-term implications are not clear.

Current research projects will, of course, expand our base of knowledge of dynamics in soil chemistry. Networks, such as the Electric Power Research Institute's Integrated Forest Study (operated through Oak Ridge National Laboratory) will be especially helpful. However, aside from the Long-Term Ecological Reserve at Coweeta Hydrologic Laboratory (funded by the National Science Foundation), there is no assured, long-term monitoring of forest soils in the South. Monitoring will be especially important for identification of the factors responsible for any observed changes in productivity.

Our Conclusions Based on Site-Specific Studies

The information from these relevant studies all indicate that:

1. Soil chemistry is dynamic; soil pH and exchangeable basic cations should be expected to change significantly on a time scale of decades.

2. Although the mechanisms that can cause soil acidification have been identified, their relative importances have not been established.
3. A monitoring network for soil chemistry and forest productivity should be established.

Previous Regional Assessments of Soil Sensitivity to Acidic Deposition

In 1972, Wiklander and Anderson published a discussion of the acidification of soils and developed sensitivity criteria for soils. They concluded that soils that would be most sensitive to acidic deposition would be sandy noncultivated soils, with current pH values above 6. Since then, many researchers have developed similar criteria for classifying soils into sensitivity groups. Five assessments have evaluated the sensitivity of soils in the South to acidic deposition: Hendrey et al. (1980) and one from Brookhaven National Laboratory by McFee (1980) as well as three successive studies from Oak Ridge National Laboratory (Klopatek et al. 1980, Olson et al.1982, and Turner et al. 1986). In this section, we discuss and evaluate the approaches used by each study, the conclusions reached by the authors, and the degree of similarity in their conclusions. A discussion of criteria of soil sensitivity follows in the next section.

Hendrey et al. 1980

Hendrey et al. (1980) used maps of bedrock geology to delineate classes of varying sensitivity to acidification, and they verified their classification in some areas by evaluating the alkalinity in waters draining those areas. Their definition of acidification focused on the export of alkalinity from soils to aquatic ecosystems. They defined four classes of bedrock. Type I provides little or no neutralization of strong acid inputs and includes granites, gneisses, and quartz sandstones. Widespread acidification from acidic deposition would be expected in Type I areas. The second type has medium-to-low resistance to acidification and consists of sandstones, shales, and conglomerates as well as some other rock types lacking carbonates. Types II areas might show impacts of acidic deposition in headwater areas, but large bodies of water would probably retain alkalinity. Type III areas are relatively resistant to drainage water acidification owing to the presence of signification amounts of calcareous rocks, mafic-to-ultramafic rocks (such as basalts) or glassy volcanic rocks. Type IV rocks contain high amounts of carbonates, such as calcareous and dolomitic limestones. No impacts of acidic deposition are expected on Type III and Type IV bedrock areas.

Many areas in the eastern United States were classed as having large proportions of Type I (sensitive) and Type II (moderately sensitive) soils. For example, Wake County in North Carolina (which includes Raleigh) had 20% Type I areas and 60% Type II areas (fig. 5.1). Neighboring Franklin County had 50% Type I and 30% Type II; however, that county also reported that alkalinity of waters exceeded 200 $\mu mol_c/L$ and appeared well buffered relative to acidic deposition. The agreement was better in other areas (particularly parts of the Blue Ridge

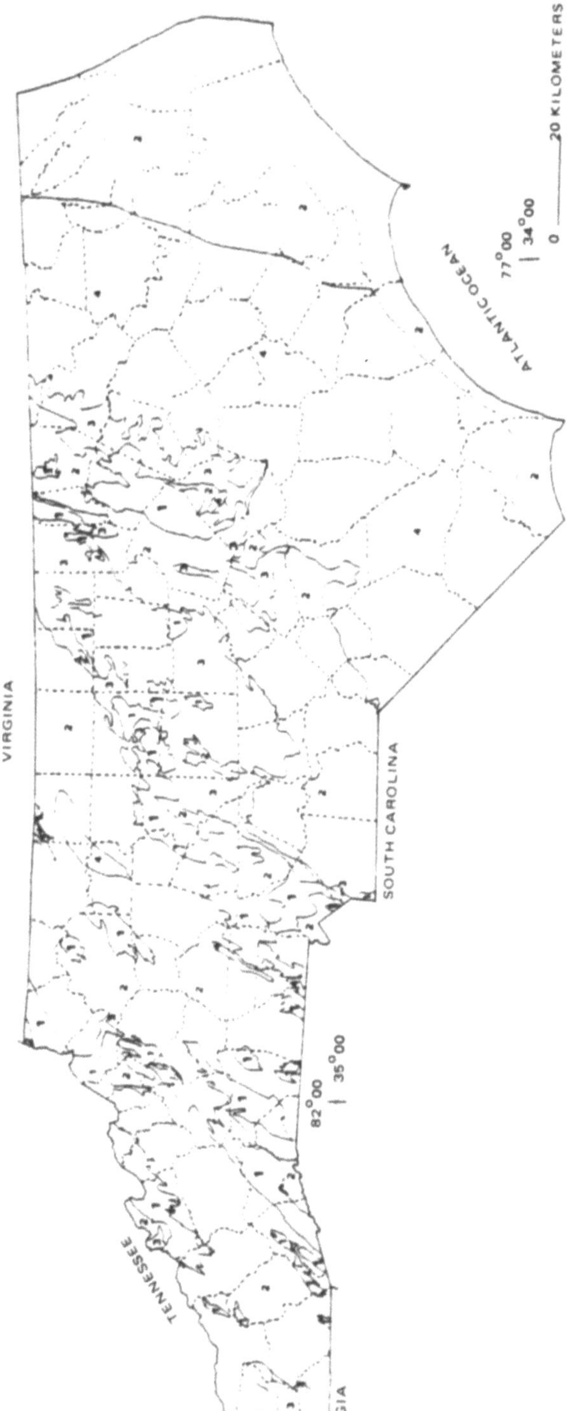

Figure 5.1. Sensitivity map of North Carolina based on geology. Highest sensitivity to acidification = 1, lowest = 4. (From Hendrey et al. 1980, reprinted by permission of G. Hendrey.)

Mountains, where counties with Type I and Type II bedrocks had alkalinities below 100 $\mu mol_c/L$.

The study by Hendrey et al. (1980) focused on sensitivity of aquatic ecosystems to acidic deposition rather than soil sensitivity. Note that the Walker Branch soil discussed earlier developed on a dolomitic parent material that would be classed by Hendrey and others as Type IV, with infinite resistance to acidic deposition. Indeed, the export of alkalinity from the Walker Branch watersheds is high (about 100 $\mu mol_c/L$, D.W. Johnson et al. 1985b). The soils (to a depth of 60 cm), however, appeared fairly sensitive to changes on a scale of decades, illustrating that classification of sensitive soils may not coincide with a classification of sensitive waters. The forest at Walker Branch may be able to utilize the reservoirs of basic cations deeper in the profile, and this largely unquantified flux of nutrients from the subsoils of Ultisols may be very important (D.W. Johnson et al. 1987).

McFee 1980

McFee's (1980) objective was to develop maps of the eastern United States that delineated soil areas by sensitivity classes. He was interested in effects of acidic deposition on both soils and aquatic ecosystems and considered all three definitions of acidification (pH, cation distributions, and alkalinity). He reviewed the previous literature and concluded that four parameters would be important in his assessment:

1. Total buffering capacity (or CEC).
2. Base saturation of the CEC (can be estimated from pH).
3. Management of the soil, including fertilization and flooding.
4. The occurrence of carbonates.

Most earlier studies had suggested that sandy soils low in organic matter and with low CEC would be most susceptible, but quantitative definitions were not available. Therefore, McFee arbitrarily chose a 25-yr time frame and assumed that if the worst-case rate of H^+ deposition (he used 100 cm/yr at pH 3.7, which equals 2 kmol H^+/ha annually) equaled 10% to 25% of the CEC in the top 25 cm of soil, a significant effect might occur. Assuming a bulk density of 1.3 kg/L, soils that ranged between about 60 to 150 $mmol_c/kg$ CEC would be classed as slightly sensitive, whereas those with less than 60 $mmol_c/kg$ would be sensitive. McFee was aware that other criteria could be used but felt his choices would delineate areas that should differ in response to acidic deposition.

McFee used a variety of soil maps, including state soil association maps, regional association maps, and some county-level maps. This information on distribution of soil types was coupled with data from the Soil Conservation Service, USDA, on soil chemistry in order to classify associations into sensitivity categories that were then mapped.

In the South, McFee concluded that the Cumberland Plateau region of Tennessee and Kentucky might be slightly sensitive and sensitive because the soils tend to be old, shallow, and formed from sandstone. Across the Coastal Plain,

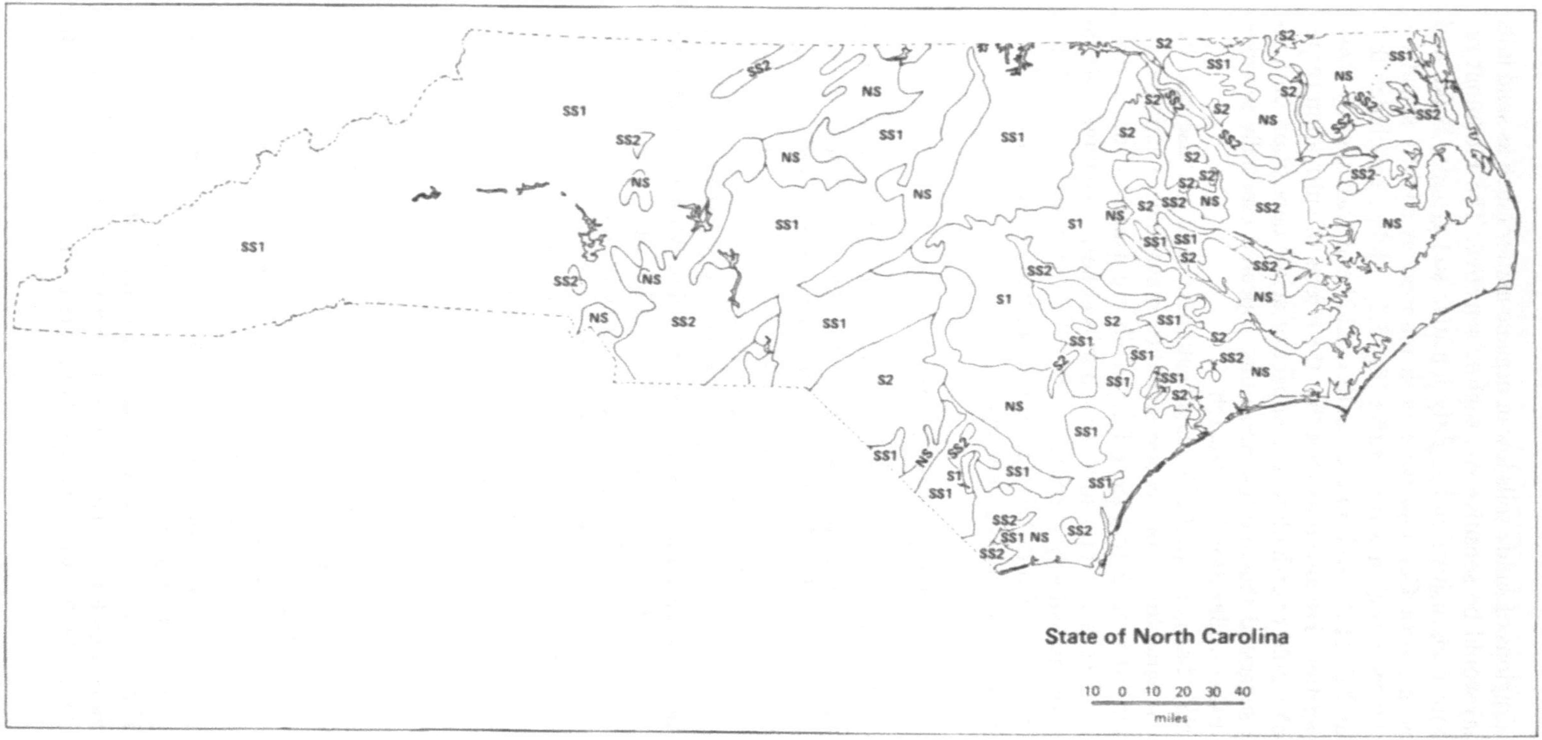

Figure 5.2. Sensitivity map of North Carolina based on soil survey data: NS = not sensitive; SS = slightly sensitive; S = sensitive; 1 = that class is dominant; 2 = that class is significant but comprises <50% of the area. (Reproduced from McFee 1980, with permission of Butterworth Publishers.)

he concluded that regions of sandy soils low in organic matter (such as sand hills in North Carolina) would be sensitive soil slightly sensitive. The Piedmont region contained both nonsensitive and slightly sensitive soil groups; Wake and Franklin counties in North Carolina were both mapped as slightly sensitive. Most of the Appalachian region was sensitive (fig. 5.2). McFee concluded that perhaps more than half of the soils in the South would be at least slightly sensitive to acidic deposition. He also indicated that the effects of cultivation (especially fertilization) would overshadow any impact of acidic deposition on agricultural soils. He suggested that the few areas that appeared especially sensitive should be given first consideration for research.

Tabatabai (1985) dismissed McFee's classification scheme as "useless" because the acidity of agricultural soils is dominated by management activities (as McFee noted), and "It is clear that since forest soils are usually acidic in reaction, measureable effects are not possible in such soils either." We think Tabatabai's presumption of insensitivity of acidic forest soils is not warranted and that the dynamics of the acidic South Carolina soil described in chapter 2 and our simulations in chapter 7 provide good evidence.

Klopatek et al. 1980

Klopatek and associates (1980) took advantage of the Geoecology Data Base developed at Oak Ridge National Laboratory to perform an evaluation very similar to that of McFee. Soil information in this data set is based on typical pedons from most of the soil great groups (such as Hapludults or Paleaqualfs) in the eastern United States, mapped at a scale of the average soil great group in each county. This level of resolution was coarser than most of McFee's work. Klopatek and others considered soil pH and exchangeable cation distributions as indexes of acidification. They noted that CEC could be used as an index of a soil's capacity to buffer acids, that base saturation could index the degree to which pedogenic processes have already consumed this buffering capacity, and that pH could index the ability of deposited H^+ to displace basic cations and acidify the soil. They used these three parameters to classify A horizons (generally top 20 to 25 cm of soil). Four classes of sensitivity were defined, based on apparently arbitrary combinations of these parameters taken in part from Wiklander 1974. Wiklander did not give CEC values, and it is unclear where Klopatek and others obtained these parameters. The most sensitive class had CEC below 120 $mmol_c$/kg, 30% to 50% base saturation, pH > 5, and a sandy texture. This matches McFee's "slightly sensitive" class. The next-most-sensitive class would have similar properties, except for a clay texture and CEC up to 200 $mmol_c$/kg. Less sensitive soils were similar to that group, except that the base saturation and pH were lower. Although these soils would be more resistant to further acidification, the authors noted that they would also provide less buffering for aquatic systems. The least sensitive soils had high CEC, high base saturation, and high pH owing to the presence of carbonate minerals.

These researchers found that on a county scale, only 117 of the 1572 counties in the eastern United States were classed as sensitive based on a CEC < 120

$mmol_c$/kg. Only one county had a CEC of < 70 $mmol_c$/kg. Using the same CEC criteria, McFee concluded that about half the soils in the South would be sensitive. McFee concluded that over half of North Carolina's soils would be slightly sensitive or sensitive, whereas Klopatek et al. concluded no counties in North Carolina had sensitive soils. Three explanations may account for the discrepancy. The inclusion of pH and base saturation criteria by Klopatek et al. may have prevented low-CEC soils from being classified as sensitive. Alternatively, the county-scale resolution used by Klopatek et al. may have masked substantial portions of sensitive soils within counties where nonsensitive soils comprised more than half the total. Finally, the extent and quality of the data sets may have differed; the modal soil great group in each county was biased toward agricultural soils in the data set of Klopatek et al.

Olson et al. 1982

Olson et al. (1982) combined the approaches of Hendrey et al. (1980) and Klopatek et al. (1980), and they included subjective evaluations of the likely importance of sulfate adsorption. County classification was based on average county conditions; insensitive counties might include a minority of sensitive soils. Their study included three stages of analysis; each stage used various criteria for defining sensitivity of soils (and aquatic ecosystems).

The first stage used a clustering algorithm to group predominantly forested counties with similar soil-buffering capacities. Highly sensitive soils were classified as having < 60 $mmol_c$/kg of exchangeable basic cations, pH below 5.5, high organic matter or low-sulfate-adsorption capacity (based subjectively on soil order), and noncarbonate rocks. This stage concluded that 55% of the nonagricultural counties in the eastern United States (east of the Mississippi River and all of Minnesota) were potentially sensitive to acidic deposition.

The second approach also included agricultural soils and used fewer soil parameters (only exchangeable basic cations and an estimate of sulfate adsorption capacity). Sensitive areas were defined as having less than 60 $kmol_c$/ha of exchangeable basic cations; sensitive areas were further subdivided into three classes, based on presumed adsorption capacities. This analysis produced results similar to the first stage, with up to 46% of all counties classified as being moderately to highly sensitive to acidic deposition. Another 38% would have been considered sensitive but were omitted due to anticipated sulfate adsorption capacity of the soil.

The third stage focused primarily on forest and range soils; sensitivity evaluations were based on soil pH (>5.5), CEC (<60 $kmol_c$/kg; there is some inconsistency in the report as to whether this represents CEC or exchangeable basic cations), and presumed sulfate adsorption capacity. Additional weightings for aquatic effects were based on bedrock type and topography. This analysis concluded that only 1% of all (nonagricultural) conunties had highly sensitive soils, with perhaps an additional 4% classified as moderately sensitive.

The differences in approach among the three stages of Olson et al. (1980) are not completely clear, and it is surprising that the third stage could conclude that only a small percentage of the counties have predominantly sensitive soils when the first two stages (using similar criteria) estimated that perhaps half of all counties were potentially sensitive. The authors expected the third stage to be more reasonable than the first two stages (only third-stage findings were reported in their abstract), but the reasons for their preference are not clear.

The inclusion of sulfate adsorption in sensitivity criteria is unique in this study; earlier assessments did not include it, and the later Oak Ridge work (Turner et al. 1986) concluded that too little information was available to provide realistic evaluations.

Turner et al. 1986

Turner et al. (1986) took advantage of improved data bases and geographic information systems and produced a detailed evaluation of soil sensitivity in the eastern United States. Their objectives were to:

1. Review and evaluate the potential for forest soil changes caused by acidic deposition.
2. Map the soils with characteristics defined as sensitive.
3. Critique the validity and usefulness of such maps.

They defined sensitive soils as those that may undergo changes because of acidic deposition, potentially leading to a reduction in forest productivity. They expected that soil-mediated effects on productivity would result from increased leaching losses of nutrient cations leading to nutrient deficiency or from increased solubility of toxic aluminum. Both of these hypotheses focus on changes in the distribution of cations on the exchange complex resulting from acidification. They established sensitivity criteria for each of these hypotheses and used a variety of data sources. Their geographic information system then allowed maps to be constructed identifying soils with overlapping values for a range of parameters. Soils data were obtained from the 1982 National Resource Inventory of the Soil Conservation Service (SCS), from the SCS Soils-5 data base, from the SCS National Pedon Data Base, from some state-level soils maps, and from the Oak Ridge National Laboratory's Geoecology Data Base.

For the base cation-leaching hypotheses, Turner et al. used sensitivity levels of: CEC (pH 7) $<$ 150 $mmol_c$/kg, base saturation 20% to 60%; and pH $>$4.5. By these criteria, 51% of forested soils in the southeast would be expected to show growth reductions owing to basic cation depletion (tables 5.1, 5.2). These criteria are very similar to those used by Klopatek et al. (1980) and by Olson et al. (1982). The data base used by Turner et al. is probably the most extensive and quality-assured set available, so we expect their more recent assessment is more accurate than earlier work at Oak Ridge National Laboratory.

For the aluminum-toxicity hypothesis, they defined sensitivity levels of: base saturation (pH 7) $<$ 20%, pH $<$ 4.5, and organic carbon $<$ 5%. In the South, 10% of forest soils met these criteria, and would be expected to show decreased forest growth because of increased aluminum toxicity under the influence of acidic deposition (tables 5.3, 5.4). Note that the ranges of base saturation do not

Table 5.1. Sensitive areas in the South with respect to basic cation-leaching hypothesis: CEC less than 150 $mmol_c/kg$, base saturation 20% to 60%, pH greater than 4.5

Forest Type	Sensitive Area (ha)	Forest-Type Area (ha)	Percentage of Forest Type	Percentage of Total Forest	Percentage of Total Region
White-Red-Jack Pine	40,100	50,100	80.0	0.1	0.1
Spruce-Fir	2,600	6,200	41.9	0.0	0.0
Longleaf-Slash Pine	2,927,700	5,469,300	53.5	6.9	3.5
Loblolly-Shortleaf Pine	6,525,500	10,804,900	60.4	15.3	7.9
Oak-Pine	6,478,700	10,612,900	61.1	15.2	7.8
Oak-Hickory	4,847,300	9,324,000	52.0	11.4	5.8
Oak-Gum-Cypress	329,100	4,920,500	6.7	0.8	0.4
Elm-Ash-Cottonwood	18,500	244,200	7.6	0.0	0.0
Maple-Beech-Birch	17,000	40,400	42.1	0.0	0.0
Aspen-Birch	900	8,600	10.5	0.0	0.0
Low Productivity	18,500	92,400	20.0	0.0	0.0
Tropical	1,300	12,300	10.6	0.0	0.0
Monstocked	530,200	1,057,200	50.2	1.2	0.6
	21,737,400	42,643,000		50.9	26.1

From Turner, et al. 1986, reprinted by permission of R. Turner.

Table 5.2. "Base" cation-sensitive areas from table 5.1, by soil taxonomic groups

Classification	Sensitive Area (ha)
Typic Glossaqualf	208,400
Typic Quartzipsamment	1,288,800
Typic Dystrochrept	466,300
Typic Haplohumod	76,800
Typic Fragiaquult	28,900
Typic Fragiudult	446,600
Typic Hapludult	11,149,600
Typic Rhodudult	470,600
Albic Glossic Natraqualf	35,100
Alfic Arenic Haplaquod	8,200
Aquic Fragiudalf	37,100
Aquic Quartzipsamment	248,900
Aquic Hapludult	999,800
Aquic Paleudult	463,300
Aquic Arenic Paleudult	143,800
Arenic Glossaqualf	58,800
Arenic Haplaquod	68,000
Arenic Haplohumod	20,200
Arenic Paleudult	506,600
Arenic Plinthaquic Paleudult	185,600
Arenic Ultic Haplaquod	3,200
Arenic Ultic Haplohumod	3,400
Entic Haplaquod	15,300
Entic Haplohumod	24,900
Fragiaquic Paleudult	198,300
Fragic Paleudult	3,300
Glossic Fragiudult	82,500
Grossarenic Haplaquod	49,600
Grossarenic Paleudult	1,598,900
Grossarenic Entic Haplohumod	56,300
Grossarenic Plinthic Paleudult	79,200
Haplaquodic Quartzipsamment	49,500
Humic Hapludult	1,000
Lithic Hapludult	40,000
Ochreptic Hapludult	205,600
Plinthic Paleudult	579,200
Psammentic Hapludult	104,600
Psammentic Paleudult	188,500
Rhodic Paleudult	362,800
Ruptic-Lithic-Entic Hapludult	29,800
Spodic Quartzipsamment	88,100
Spodic Paleudult	14,400
Ultic Haplaquod	607,900
Ultic Haplohumod	7,700
Vertic Paleudalf	413,900
Vertic Hapludult	18,100
	21,737,400

From Turner et al. 1986, reprinted by permission of R. Turner.

Table 5.3. Sensitive areas in the South with respect to aluminum-toxicity hypothesis: base saturation less than 20%, pH less than 4.5, organic carbon less than 5%

Forest Type	Sensitive Area (ha)	Forest-Type Area (ha)	Percentage of Forest Type	Percentage of Total Forest	Percentage of Total Region
White-Red-Jack Pine	13,00	50,100	2.6	0.0	0.0
Spruce-Fir	3,200	6,200	51.6	0.0	0.0
Longleaf-Slash Pine	1,173,800	5,469,300	21.5	2.8	1.4
Loblolly-Shortleaf Pine	820,700	10,804,900	7.6	1.9	1.0
Oak-Pine	801,700	10,612,900	7.6	1.9	1.0
Oak-Hickory	442,300	9,324,000	4.7	1.0	0.5
Oak-Gum-Cypress	604,600	4,920,500	12.3	1.4	0.7
Elm-Ash-Cottonwood	7,000	244,200	2.9	0.0	0.0
Low Productivity	10,900	92,400	11.8	0.0	0.0
Monstocked	150,900	1,057,200	14.3	0.4	0.2
	4,016,400	42,581,700		9.4	4.8

From Turner et al. 1986, reprinted by permission of R. Turner.

Table 5.4. Aluminum-toxicity sensitivity[a] areas from table 5.3, by soil taxonomic group

Classification	Sensitive Area (ha)
Typic Paleaquult	1,190,000
Aeric Haplaquod	699,400
Aeric Paleaquult	408,400
Arenic Paleaquult	657,700
Arenic Plinthic Paleaquult	187,600
Arenic Umbric Paleaquult	278,400
Grossarenic Paleaquult	319,100
Plinthic Paleaquult	70,700
Umbric Paleaquult	205,100
	4,016,400

[a] Base saturation < 20%, pH < 4.5, organic carbon < 5%.
From Turner et al. 1986, reprinted by permission of R. Turner.

overlap between the two hypotheses, so this 10% would be added to the 50% from the nutrient-depletion hypothesis to indicate that 60% of southern forest soils are at risk with regard to either nutrient deficiency or aluminum toxicity.

This study leads to several clearly testable hypotheses. For example, their conclusions predict soils that fall within the class of sensitivity by the nutrient depletion hypothesis should respond to fertilization with nutrient cations (especially soils towards the most sensitive tail of the class). Similarly, liming to reduce aluminum availability should stimulate tree growth on soils that fall within the aluminum-sensitivity class.

Our Conclusions from the Regional Assessments

There were striking differences among the four studies, despite similar choices of sensitivity criteria. Klopatek et al. (1980), and one of the three stages of Olson et al. (1982) indicate that only a very small fraction of counties in the eastern United States are likely to be sensitive to acidic deposition. The other studies conclude that perhaps half of the eastern United States is likely to be at least moderately sensitive. Much of the disparity results from the use of different data bases; the work by Turner et al. (1986) was based on the best data set. The selection of sensitivity criteria was fairly subjective, but surprisingly uniform in all studies. The loblolly pine soil in the 20-yr South Carolina study (chap. 2) would have been classified as sensitive by all the schemes, except for the third stage of Olson et al. (1982), consistent with the rapid acidification observed.

Critique of Criteria for Gauging Soil Sensitivity

The sensitivity of soils to acidic deposition can be quantified using three approaches: as a decrease in soil pH, through changes in the concentration and distribution of cations on the soil exchange complex and in the soil solution, and as a decrease in the alkalinity of the soil solution. The sensitivity index used varies with the availability of data as well as the objectives of any assessment or classification exercise.

Criteria Used in Previous Assessments

The regional assessments described earlier were fairly consistent in their sensitivity criteria. McFee (1980) chose a 25-yr time frame, assumed a worse-case scenario of 2 kmol H^+/ha annual deposition with the cation exchange complex providing the only source of H^+ buffering or neutralization. If the total deposition for 25 yr equaled 10% to 25% of the exchange complex in the top 25 cm, the soil was classified as sensitive. By this definition, soils with less than about 60 $mmol_c$/ha of CEC were considered sensitive. The limitations of this definition include:

1. The deposition rate of 2 kmol H^+/ha greatly exceeds current input rates, and reduction of sulfate and nitrate within ecosystems neutralizes much of the desposited H^+ (Binkley and Richter 1987). We expect 0.5 to 1.0 kmol H^+/ha would be more reasonable.
2. The soil data available to McFee included CEC measured at pH 7. This parameter substantially overestimates the CEC that is effective under ambient conditions in acidic soils. The actual CEC of acidic soils would be in the range of 10% to 50% of the values McFee used. Further, the ability of the exchange complex to buffer inputs of H^+ depends primarily on the degree of dissociation of the complex (i.e., the exchangeable basic cations) and not on the total capacity.
3. Therefore, as McFee noted, inclusion of basic cation contents of soils would have improved the assessment.

Klopatek et al. (1980) and Olson et al. (1982) used criteria of sensitivity derived from Wiklander (1974, 1980). The latter chose these criteria as indexes of buffering capacity against pH change, ability to retain H^+ deposited from the atmosphere or generated within the ecosystem, and the efficiency with which H^+ can displace basic cations from the exchange complex. Wiklander's choice of sensitivity criteria was based on laboratory experiments of the ability of added H^+ to replace exchangeable basic cations. He did not, however, select a critical value for CEC, and it is unclear where the two Oak Ridge studies obtained their values.

The assumption that maximum sensitivity to H^+ input will occur in soils with low CEC and moderate base saturation was challenged by Reuss and Johnson (1986). They pointed out that the rate of leaching of basic cations might be maximized under those conditions but that the rate of change in soil pH would be greater if the base saturation were already very low (see fig. 2.2). This pattern is analogous to the buffer range of weak acid; the maximum buffering (minimum change in pH) occurs when the acid is half dissociated, and the greatest changes in pH occur when the acid is almost completely protonated or dissociated. In addition, the concentration of aluminum ions in soil solution increases dramatically with moderate changes in base saturation when base saturation is low, perhaps less than 20% (Reuss and Johnson 1986). The assessment by Turner et al. (1986) improved on earlier work by considering soils as sensitive to basic cation depletion if they had moderate base saturation, or sensitive to aluminum mobilization if they had low base saturation.

It is important to note that the exchange complex represents only part of the buffer capacity of soils (Ulrich 1987, Magdoff et al. 1987). As an illustration, the soil titration curve for the South Carolina pine soil in 1962 is compared with the expected buffer curve if exchangeable basic cations provided all the ANC (fig. 5.3). Before the exchangeable basic cations are completely depleted, other mechanisms, such as dissolution of aluminum hydroxides (Magdoff et al. 1987, Mulder et al. 1987) contribute a substantial amount of ANC.

Only the assessments by Olson et al. (1982) included sulfate adsorption capacity; unfortunately, they had no survey data on sulfate adsorption and had to estimate relative capacities based on soil orders. Ultisols were thought to have high capacity, Inceptisols moderate capacity, and Spodosols low capacity. More recent work has shown these expectations may be wrong: D.W. Johnson et al. (1986) reported that two Inceptisols in the state of Washington retained large amounts of added sulfate, but that two Ultisols in Tennessee retained none. D.W. Johnson (pers. comm. 1987) concludes that earlier generalizations based on soil orders are likely wrong and that more direct data on sulfate adsorption capacity are needed.

New Recommended Criteria

As discussed in chapter 1, root activity of trees is relatively insensitive to direct variations in the soil/soil solution pH. However, pH is a master variable that often regulates solute activity and changes in pH may have profound indirect effects on forest biota. The indirect effects that we feel are important involve:

1. Cation exchange capacity near ambient pH.
2. Extractable basic cations.
3. Extractable soil aluminum.
4. Sulfate adsorption.
5. Cation selectivity coefficients.
6. The effect of items 1 to 5 on soil-solution chemistry, particularly the concentrations of cation nutrients and aluminum.
7. The alkalinity of drainage waters.

Ideally, we would like to establish sensitivity criteria (defined as levels where potentially harmful effects need to be considered) and then test them with simulation modeling and field experiments. However, critical levels of these parameters are difficult to determine, thus for some of them, our best estimates come from the dynamic simulations.

For CEC, we cannot derive a critical level based on available empirical information. Therefore, we can only investigate the importance of CEC in our simulation modeling (chap. 7), and suggest validation experiments.

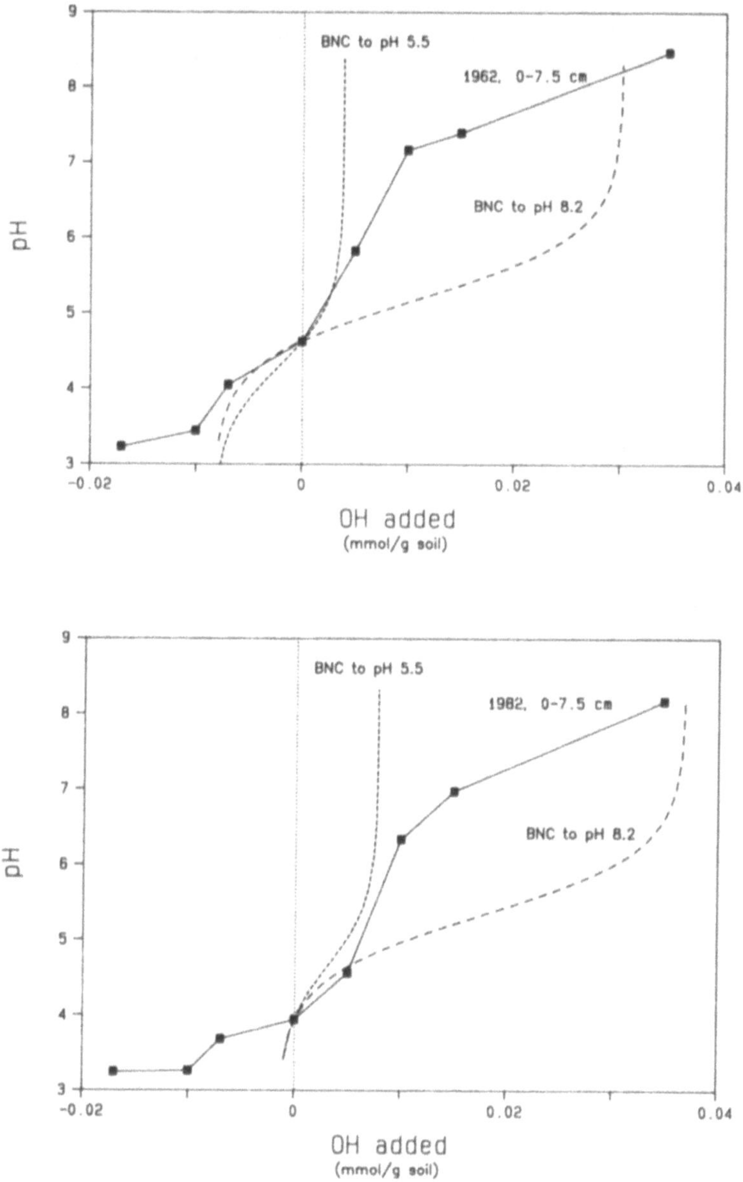

Figure 5.3. Titration curves (solid lines) for the 1962 (*top*) and 1982 (*bottom*) pine plantation soils described in chapter 2 do not match the hypothetical curves based on the Henderson-Hasselbalch equation with a single pK and BNC to either 5.5 (short dashes) or 8.2 (long dashes). A single pK cannot represent the acid/base behavior of the soils at either point in time, and a substantial amount of buffer capacity at low pH (ANC) exists in addition to what is provided by the exchangeable base cations (the dashed lines represent removal of basic cations).

The critical level of basic cations for tree nutrition is also poorly known. At present, we suspect some pine productivity on some soils in the South is limited by K^+ availability. Based on the data of Turvey and Allen (1987), our best estimate of the critical concentration of exchangeable potassium might be about 0.5 $mmol_c/kg$ or 0.75 $kmol_c/ha$ (see chap. 6).

The critical concentrations for extractable soil aluminum are also difficult to determine. Growth response of loblolly pine to phosphorus fertilization appears to increase with increasing exchangeable aluminum (Wells et al. 1973, Hart et al. 1986). This trend could be due to covariance of exchangeable aluminum and N availability, or to aluminum control of P solubility (sites high in aluminum respond better to P additions because native P is bound by aluminum). Based on the work of Hart et al. (1986), soils with more than 25 $mmol_c/kg$ of Al^{3+} might have restricted availability of P. However, we believe this value would vary with soil pH, is very speculative, and should be tested more directly than in previous studies.

Sulfate adsorption isotherms are available for only a few soil types in the southeast, and these few data do not support broad generalizations among soil types (cf. D.W. Johnson et al. 1986). We have not developed an estimate of the critical value of sulfate adsorption because this process affects vegetation only through indirect effects on other processes. However, the indirect effects may be important enough that substantial research is warranted on sulfate adsorption characteristics of southern forest soils.

As with sulfate, we have little idea of the variability in cation selectivity coefficients for southern soils. It is possible that two soils could have similar chemistries, but one would be more sensitive to acidic deposition because of different tendencies for basic cations to be replaced by aluminum. We have not attempted to estimate a critical range of selectivity coefficients because their importance varies with other parameters. The importance of cation selectivity coefficients is demonstrated in the sensitivity analyses in chapter 7.

The effects of CEC and exchangeable basic cations and aluminum on soil solution is typically characterized with simulation models. Therefore, critical values for the exchange complex that lead to critical concentrations in the soil solution depend on two factors: sensitivity of plants to soil solution concentrations and accuracy of models in predicting soil-solution concentrations. Testing model accuracy is a high priority; the validity of the models will be established more clearly in the next few years. As more is learned, improved models will provide the opportunity for improved assessments. Critical concentrations of solution nutrients and aluminum on plant nutrition and vigor are harder to test, and more research focused on these questions is needed.

Solution culture studies have examined the effects of aluminum concentrations on root and plant growth. For example, researchers at the State University of New York at Syracuse have examined the sensitivity of several tree species to ranges of aluminum concentrations in solution culture (Thornton et al. 1986a, 1986b, 1987). They found that growth of honeylocust (*Gleditsia triacanthos* L.) seedlings was reduced at aluminum concentrations of 20 $\mu mol/L$ (60 $\mu mol_c/L$). Growth of red spruce seedlings decreased at 100 to 200 $\mu mol/L$ for red spruce, consistent with the reductions observed by Joslin (1987) and Hutchinson et al.

(1986) for red spruce, black spruce [*P. mariana* (Mill.) B.S.P.] and white spruce [*P. glauca* (Moench) Voss]. Loblolly pine was more resistant, showing no growth reduction below 300 to 1000 μmol/L (depending on genotype). In contrast, Paganelli et al. (1987) found that the number and length of new roots on loblolly pine seedlings declined as Al levels exceeded 185 μmol/L. Northern red oak (*Quercus rubra* L.) showed growth declines only above 500 μmol/L. Hutchinson et al. (1986) found that seedlings of jack pine (*P. banksiana* Lamb.) and white pine (*P. strobus* L.) showed growth declines only at aluminum concentrations exceeding 700 μmol/L. These researchers also noted that the uptake of nutrient cations was diminished somewhat at aluminum concentrations below those required to reduce growth. The ratio of Ca/Al and Mg/Al may be more important than the aluminum concentration per se (Brown 1983, Huag 1984). For example, Rost-Siebert (1983) concluded that Norway spruce roots are likely to be damaged at Ca/Al (molar basis) decreased below 1:1. Meyer et al. (1985) found that the number of live root tips in mature Norway spruce stands was reduced when the Mg/Al ratio decreased below 0.9 Mg/1.0 Al.

Extrapolation of solution culture studies to forest soil situations is difficult. For example, Matzner et al. (1986) tested the hypothesis that low soil pH (and high aluminum concentrations) inhibited growth of Norway spruce roots by examining root colonization of soil cores buried in mesh bags. Root colonization of limed cores was greater than in control cores; unfortunately, the response could not be apportioned to either reduced aluminum concentrations or increased calcium and magnesium concentrations because of the confounded experimental design.

Kelly et al. (1987) grew red spruce seedlings in forest soils amended with aluminum. They found that the concentrations of aluminum in soil solution correlated highly ($r^2 = 0.78$) with biomass production; the critical level appeared to be about 100 to 300 μmol Al/L, very similar to findings from solution cultures.

Assessments of aluminum toxicity to plants growing in forest soils are also complicated by the presence of complexing ligands in natural soil solutions as well as varying responses to different aluminum species. A number of investigators have studied the distribution of aluminum in acidic soil solutions (David and Driscoll 1984, Driscoll et al. 1985, Cronan et al. 1986), though unfortunately no studies are currently available for the South. In general, solutions draining organic horizons show high concentrations of aluminum, but this aluminum is largely complexed with naturally occurring organic solutes. As the solutions migrate through the mineral soil, concentrations of dissolved organic carbon and organic Al decrease while inorganic forms of aqueous aluminum increase. Inorganic aluminum largely originates from the mineral soil, with the concentration and speciation varying with the pH of the soil as well as with concentrations of inorganic complexing ligands (F^-, SO_4^{2-}). Under highly acidic conditions ($pH < 4.5$), inorganic aluminum is largely in an aquo (Al^{3+}) form because the concentration of aqueous aluminum greatly exceeds the concentrations of available ligands. With increases in pH, concentrations of inorganic aluminum decrease. As the concentrations of inorganic aluminum decrease, the relative distribution of inorganic complexes involving fluoride and sulfate increases (Driscoll and Schecher 1988).

The sensitivity of forest vegetation to different aluminum species is not well established. Some information is available from the agronomic literature. For example, in carefully controlled experiments, Pavan et al. (1982) found that aluminum toxicity to coffee was closely coupled to the activity of aquo-aluminum. In a recent investigation, D.R. Parker et al. (1987) found that wheat was sensitive to elevated concentrations of both aquo-aluminum and polymeric forms of aluminum. In fact, polymeric forms were more toxic than aquo-aluminum. Toxicity was not evident when aluminum was complexed by either organic or inorganic ligands.

Given the apparent variability in plant response, concentrations of total aluminum or ratios of aluminum to nutrient cations may not be good measures of toxicity to forest vegetation. With these limitations in mind, our current best estimate for critical concentrations of aluminum in soil solution are: < 10 μmol/L, probably no effect; 10 to 50 μmol/L warrants consideration for sensitive species; 50 to 250 μmol/L warrants consideration for moderately sensitive species. Lacking better data, we tentatively accept the German estimates that solution concentration ratios of less than 1:1 Ca/Al or Ca/Mg may impair roots.

Most forest soil solutions have much lower aluminum concentrations, and much higher Ca/Al ratios than these suggested critical concentrations. However, our computer simulations suggested some forest soils in the South might approach these levels, and Cronan and Goldstein (1987) reported concentrations of aluminum in forest soil solutions of up to 240 μmol/L in the ALBIOS study funded by the Electric Power Research Institute.

The final parameter of sensitivity to acidification, the alkalinity of soil solutions, relates more to possible impacts on aquatic ecosystems than to soils. The degree of dissociation of carbonic acids in forest soils is very sensitive to changes in pH. When pH decreases from 5.2 to 4.9, bicarbonate protonation results in a marked decline in solution alkalinity (Reuss and Johnson 1986). At Coweeta, an estimated decrease in soil pH of only 0.1 to 0.15 units appeared to decrease export of alkalinity by 30%. This process has little impact on the soil and might even retard leaching losses of nutrient cations. However, the export of bicarbonate from forest soils to aquatic ecosystems represents a major buffering mechanism, and a large reduction in alkalinity may allow a fairly large decrease in lake pH (and increase in aluminum concentrations). Streams and lakes containing below 40 or 50 μmol/L of alkalinity are generally considered to be poorly buffered (Colquhoun et al. 1984). Some acidic soils export water containing less than this amount; in our simulation runs (chap. 7), if acidic deposition caused alkalinity in a soil solution to decrease from substantially above 50 $\mu mol_c/L$ to substantially below this value, we concluded that acidic deposition may acidify that soil with regard to its ability to provide buffering capacity to aquatic ecosystems.

Improvements in the Definitions of Sensitive Soils

The sensitivity criteria also need to be tested in the field. For example, growth responses to fertilization with basic cations need to be tested on soils that are classified as sensitive by current standards. Soils that are classified as sensitive by aluminum toxicity hypotheses should be limed (with sodium carbonate, hydroxide, or oxide to prevent confounding of calcium or magnesium effects on tree nutrition) to verify vegetation response. Chapter 6 considers such experimental evidence that is already available from fertilization trials, and chapter 7 examines the sensitivity of simulation model predictions to the criteria of soil sensitivity used in these regional assessments.

6. Nutrient Cycles and Nutrient Limitations in the South

The impact of acidic deposition on forest soils depends largely on the nutrient cycles of the forest, and management treatments (such as fertilization) may also be important. As noted in chapter 2, if a forest utilizes the nitrate or sulfate from atmospheric deposition, the associated acidity is neutralized. The acidity of a soil with an abundant supply of nutrient cations may be less affected than a similar soil with a lower supply, and soil changes may not be proportional to changes in forest productivity. In this chapter, we review the cycles of nitrogen, sulfur, and cations and we discuss the degree to which these nutrients limit forest productivity in the South.

The Nitrogen Cycle

As discussed in chapter 2, oxidation and reduction reactions are important in the nitrogen cycle. The biochemical electron-transferring reactions strongly influence the availability of nitrogen to plants and the net impact of nitric acid deposition. The atmosphere is comprised of 78% N_2, but this triple-bonded molecule is difficult to oxidize or reduce, and it is unavailable to most plants. Only plants with symbiotic nitrogen-fixing procaryotes (such as legumes with the bacteria *Rhizobia* or a few nonlegumes with the actinomycete *Frankia*) have access to the N_2 pool. These procaryotes have an enzyme, nitrogenase, that reduces N_2 to form ammonia (NH_3), which is readily utilized by plants.

Few studies have examined nitrogen fixation in the South. In the absence of symbiotic nitrogen fixers, rates are probably less than 0.07 kmol-N/ha (1 kg-N/ha) annually in most forests (for a summary, see Grant and Binkley 1987). One study with the symbiotic N fixer, wax myrtle (*Myrica cerifera*), estimated that about 0.7 kg-N/ha (10 kg-N/ha) was fixed annually in a slash pine forest in Florida (Permar and Fisher 1983). Where black locust dominates, N-fixation rates may be on the order of 3.5 to 10.5 kmol-N/ha (50 to 150 kg-N/ha) annually (Boring and Swank 1984).

Once nitrogen has been incorporated into plant tissues, it recycles within the plant or ecosystem until lost through leaching, fire, dentrification, or harvest. This recycling involves decomposition of plant and microbial tissues. Decomposer microbes secrete enzymes that attack the chemical bonds in litter, and the microbes then absorb small organic molecules. Inorganic nitrogen compounds are released either as by-products of microbial scavenging for energy or by enzymes that specifically sever C—N bonds. This process is termed mineralization, and inorganic molecules containing nitrogen are referred to as either mineral or inorganic forms. Nitrogen does not occur in rocks in significant concentrations.

The first product of mineralization is ammonia, which immediately adds a H^+ to become ammonium (except when pH exceeds 8). This protonation means that the soil is a stronger acid than ammonium; the soil acid complex dissociates and the ammonia protonates. Ammonium is also deposited from the atmosphere, at rates from about 0.1 $kmol_c$/ha in the South to several $kmol_c$/ha in parts of Europe. Ammonium may be stored on cation exchange sites before it is taken up by plants and processed back into proteins. It may also be utilized (immobilized) by microbes. Measurements of mineralization rates usually cannot determine how much of the mineralized nitrogen was immobilized by microbes (it may be more than 95%), so such measurements are usually termed net mineralization (see Binkley and Hart 1989).

Ammonium is a high-energy compound that some microbes utilize as an energy source by transferring eight electrons from ammonium to oxygen, forming water and nitric acid (termed nitrification). Nitrate also can be utilized by microbes and plants, but it takes more energy to process because it must be reduced back to the ammonia state to form proteins. This gain of electrons coincides with retention of a H^+, neutralizing the acidity formed in nitrification. Note, however, that the neutralizing step may occur at a different time or location from the H^+ generating step, allowing for within-system changes in acidity and alkalinity.

Annual estimates of nitrogen mineralization are available for a variety of forests in the South (table 6.1). Annual mineralization rates range from less than 2 kmol-N/ha (28 kg-N/ha) to more than 12 kmol-N/ha (170 kg-N/ha); the techniques used to assess mineralization varied, the contribution of the choice of method to the variation found among sites is problematic (see Binkley and Hart 1989). Most sites had annual rates between 2 and 6 kmol-N/ha (28 to 85 kg-N/ha). Little nitrate was formed, except in distured (such as harvested and site-prepared locations) and high-mineralization sites (such as black locust and beech stands).

The amount of nitrogen required to produce new (aboveground) tissues

Table 6.1. Nitrogen mineralization estimates for forests in the South

Location	Vegetation	Treatment	Method	Annual Mineralization		
				kmol-N/ha	kg-N/ha	Reference
Henderson, NC	22-yr-old loblolly pine	undisturbed	buried bags	1.6	22	Vitousek and Matson 1985
		harvested, site-prepped		6.0–8.5	85–120	
Duke Forest, NC	20- to 100-yr old loblolly pine	undistributed	buried bags	1.4–3.2	20–45	Satterson 1985
	> 150-yr-old mixed hardwoods	undisturbed		5.0	70	
Patrick County, VA	loblolly pine, mixed hardwoods	undisturbed	buried bags	2.7	38	Fox et al. 1986
		harvested, site-prepped		4.3–6.8	60–95	
Gainesville, FL	slash pine	undisturbed	core/resins	3.6–7.2	50–100	DiStefano 1984[a]
Coweeta, NC	oak-hickory	undisturbed	buried bags	1.8	25	Montagnini 1988
	pine, mixed hardwoods	undisturbed		6.4	90	
	black locust	undisturbed		8.2	115	
Clingman's Dome, NC	spruce/fir	undisturbed	core/resins	2.2–12.9	31–180	Strader et al. 1988
Mt. Mitchell, NC	spruce/fir	undisturbed	core/resins	2.4–10.4	34–146	Strader et al. 1988
Whitetop Mountain, VA	spruce/fir	undisturbed	core/resins	1.9–9.1	26–128	Strader et al. 1988

Mt. LeConte, NC	Fraser fir	mature	core/resins	7.1	100	Sasser and Binkley 1988
		dead		5.6	78	
		regenerating		2.9	40	
		juvenile		2.5	35	
Smoky Mountains, NC	beech	undisturbed	core/resins	7.0	95	Bloss and Binkley 1988
		boar-rooted		7.0	95	

[a] Extrapolated from monthly assays.

Table 6.2. Nitrogen pools (kmol/ha) and cycling rates (kmol ha^{-1} yr^{-1}) for low-elevation forests in the southeastern United States

Location, Vegetation, Age (in yr)	Biomass N	Requirement	Annual Litterfall N	Reference
Boone County, Mo, Oak species, 35–92	9.9	4.6	2.3	Rochow 1975
Norman, OK, Oak species, 80	82.6	14.0	4.6	Johnson and Risser 1974
Durham, NC, Loblolly pine, 16	18.4	4.8	4.1	Wells and Jorgensen 1975
Flatwoods, MS, loblolly pine, 5	5.1	2.4	—	Smith et al. 1977
Flatwoods, MS, Loblolly pine, 20	12.4	—	—	Switzer and Nelson 1972
Gainesville, FL, Slash pine, 26	15.7	3.3	1.6	Gholz et al. 1985

ranges from about 2 to 5 kmol-N/ha (30 to 75 kg-N/ha) for forests in the South (table 6.2). Probably one-third to one-half of this requirement is met from recycling of nitrogen from senescent foliage, so the rate of nitrogen uptake commonly ranges from 1.4 to 4.8 kmol-N ha^{-1} yr^{-1} (20 to 60 kg-N ha^{-1} yr^{-1}). Most of this nitrogen is returned to the forest floor annually as litterfall; the rate of accumulation of nitrogen in biomass is only about 0.4 to 1.4 kmol-N ha^{-1} yr^{-1} (5 to 20 kg ha^{-1} yr^{-1}). In general, rates of uptake and accumulation in young stands are probably higher than in older stands, and rates in hardwood forests may be greater than in pine forests.

The rates of mineralization of nitrogen is usually lower than the assimilation capacity of vegetation and microbes, given ample availability of other site resources. Most ecosystems in the South can be considered to be nitrogen limited, as evidenced by the low losses of N from the ecosystems listed in table 6.3. Based on studies from other regions, we expect that even when large amounts of fertilizer nitrogen are added to southern forests, the capacity of plants and microbes to absorb nitrogen is sufficient to retain most of the added nitrogen (Binkley 1986). However, few studies have examined the retention of fertilizer nitrogen in the South; some have not been able to find all of the added nitrogen within the ecosystem after one to several years (Baker et al. 1974, Mead and Pritchett 1975).

The assimilation capacity of vegetation should decline if the overall growth rate of a forest declines, so nitrogen retention capacity may be reduced in forests showing decline (from whatever mechanism). Therefore, nitrogen leaching from declining forests should be higher than for healthier forest, and such high leaching losses may reflect an effect of decline rather than cause (Hauhs and Wright 1986).

Table 6.3. Input and output (kmol$_c$/ha annually) estimates from ecosystem nutrient budgets

Location, Vegetation	SO_4^{2-}	NO_3^-	NH_4^+	Ca^{2+}	Mg^{2+}	K^+	H^+	Reference
Bradford Forest, FL, slash pine								Riekirk et al. 1978
Input	—	0.45	0.26	0.76	—	0.15	—	
Output	—	0.01	0.04	0.29	—	0.06	—	
Net	—	+0.44	+0.22	+0.47	—	+0.09	—	
Santee Watershed, SC, loblolly pine								Gilliam 1984
Input	0.50	→ 0.12←		0.26	0.13	0.03	0.5	
Output	0.47	→ 0.00←		0.52	0.19	0.03	0.0	
Net	+0.03	→+0.12←		−0.26	−0.06	0.00	+0.5	
Clemson, SC, loblolly								D.W. Johnson et al. 1985a
Input	0.77	0.06	0.22	0.14	0.08	0.06	0.5	
Output	0.44	0.00	0.00	0.07	0.08	0.02	0.0	
Net	+0.73	+0.06	+0.22	+0.07	0.00	+0.04	+0.5	
Duke Forest, NC, loblolly pine								K. Knoerr, D. Binkley, unpub. data
Input	0.67	0.17	0.07	0.10	0.04	0.09	0.2	
Output	1.40	0.00	0.00	0.59	0.30	0.20	0.0	
Net	−0.73	+0.17	+0.07	−0.49	−0.26	−0.11	+0.2	
Coweeta, NC, mixed hardwoods								Waide and Swank 1987
Input	0.67	0.22	0.15	0.23	0.08	0.05	0.6	
Output	0.09→0.44	0.01	0.00	0.30	0.28	0.11	0.0	
Net	+0.58→0.23	+0.21	+0.15	−0.07	−0.20	−0.06	+0.6	

Table 6.3. *Continued*

Location, Vegetation	SO_4^{2-}	NO_3^-	NH_4^+	Ca^{2+}	Mg^{2+}	K^+	H^+	Reference
Chesapeake Bay, MD, mixed hardwoods								Weller et al. 1986
Inputs	0.95	0.41	0.24	0.15	0.11	0.06	0.9	
Outputs	0.74	0.01	0.01	0.24	0.32	0.11	0.0	
Net	+0.21	+0.40	+0.23	−0.09	−0.21	−0.05	+0.9	
Catoctin Mountains, MD, mixed hardwoods								Katz et al. 1985
Hauver Branch								
Inputs	0.72	0.32	—	0.16	0.05	0.01	1.1	
Outputs	1.36	0.21	—	1.86	1.31	0.07	0.0	
Net	−0.64	+0.11	—	−1.70	−1.26	−0.06	+1.1	
Hunting Greek								
Inputs	0.72	0.32	—	0.16	0.05	0.01	1.1	
Outputs	1.21	0.39	—	2.50	1.44	0.22	0.0	
Net	−0.49	−0.07	—	−2.34	−1.39	−0.21	+1.1	
Soldier's Delight (Serpentinite parent material), MD, mixed hardwoods								Cleaves et al. 1974
Inputs	0.90	0.37	0.31	0.25	0.08	0.04	—	
Outputs	0.66	0.03	0.01	0.18	2.93	0.01	—	
Net	+0.24	+0.34	+0.30	+0.07	−2.85	+0.03	—	
Pond Branch (schist parent material), MD, mixed hardwoods								Cleaves et al. 1970
Inputs	0.37	—	—	0.16	0.08	0.04	—	
Outputs	0.08	—	—	0.15	0.15	0.05	—	
Net	+0.29	—	—	+0.01	−0.07	−0.01	—	

Shenandoah Park, VA, mixed hardwoods								P. Ryan, pers. comm. 1987
Input	0.56	0.23	0.13	0.12	0.04	0.02	0.5	
Output	0.19	0.19	0.00	0.05	0.08	0.08	0.0	
Net	+0.37	+0.04	+0.13	+0.07	−0.04	−0.06	+0.5	
Fernow Forest, WV mixed hardwoods								Helvey and Hunkle 1986
Input	0.83	0.43	0.55	0.40	—	—	0.8	
Output	0.53	0.61	0.05	0.62	—	—	0.0	
Net	+0.30	−0.19	+0.05	−0.22	—	—	+0.8	
Walker Branch, TN								Johnson et al. 1985a
Chestnut oak								
Input	0.92	0.14	0.09	0.35	0.08	0.05	0.7	
Output	1.22	0.00	0.00	0.27	0.28	0.06	0.0	
Net	−0.30	+0.14	+0.09	+0.08	−0.20	−0.01	+0.7	
Yellow poplar								
Input	0.92	0.14	0.09	0.35	0.08	0.05	0.7	
Output	2.28	0.00	0.00	1.49	0.47	0.10	0.0	
Net	−1.36	+0.14	+0.09	−1.14	−0.39	−0.05	+0.7	
Loblolly pine								
Input	0.82	0.09	0.05	0.32	0.06	0.03	0.7	
Output	1.30	0.00	0.01	1.09	0.32	0.13	0.0	
Net	−0.48	+0.09	+0.05	−0.77	−0.25	−0.10	+0.7	
Camp Branch, TN mixed hardwoods								Johnson et al. 1985a
Input	1.25	0.31	0.42	0.49	0.11	0.07	0.2	
Output	0.63	0.02	0.04	0.05	0.15	0.08	0.0	
Net	+0.62	+0.29	+0.38	+0.44	−0.04	−0.01	+0.2	

The production (or deposition) of nitrate is important to both the availability and retention of nitrogen in ecosystems. The negative charge of the ion leaves it unaffected by the CEC, making it more mobile and available to plants and susceptible to leaching loss. Nitrate appears to be immobilized less readily by microbes than is ammonium, which may further increase availability (Firestone 1987) and opportunities for loss. Nitrate may also be involved in denitrification ("dissimilatory reduction"), where it used as an electron acceptor in the absence of oxygen and converted to NO, N_2O, or N_2 gases. However, leaching losses of nitrate are generally small; only 2 of the 16 ecosystems listed in table 6.3 showed even marginal net losses of nitrogen. Exceptions to this generality should be expected where nitrogen mineralization is especially high, such as in black locust forests, and perhaps in some deforested areas. Denitrification is very difficult to measure accurately, but it is probably small (< 0.07 kmol N/ha annually) in most undisturbed forest in the South (Davidson and Swank 1986, Robertson et al. 1987, Wells et al. 1988). Robertson et al. (1987) concluded that denitrification in an Ultisol under loblolly pines might be limited by nitrate availability and that rates might increase to about 0.3 to 0.5 kmol N/ha annually in response to elevated nitrate concentrations after harvest and site preparation. More information on denitrification is needed for drained and bedded sites in the Lower Coastal Plain, where close proximity of aerobic and anaerobic soil horizons may favor denitrification.

The Sulfur Cycle

The cycling of sulfur is similar to nitrogen, but with several additional features. Elemental sulfur (S^0) is easily metabolized by microbes and is unimportant in natural ecosystems. The major gaseous form, sulfur dioxide (SO_2), can be converted to sulfuric acid in the atmosphere or directly absorbed and utilized by plants (Mengel and Kirkby 1985). Whereas the most reduced form of nitrogen is a weak acid that remains as an ion in soil solutions (ammonium), the most reduced form of inorganic sulfur is hydrogen sulfide (H_2S), which is lost from soils as a gas under acidic conditions or, in the presence of metals (generally iron), it may be precipitated as a mineral (such as FeS or FeS_2).

The major ionic form of sulfur found in ecosystems is the completely oxidized form of sulfate (SO_4^{2-}). Once taken up by plants or microbes, it is reduced before incorporation into organic molecules. This reduction consumes electrons and H^+, so assimilation of sulfate neutralizes acidity just as with nitrate. A small amount of sulfur may remain as sulfate, bound to organic molecules by an ester linkage (C—O—S bond).

As with nitrogen, sulfur is released either as by-product of microbial scavenging for energy or by enzymes that cleave C—O—S bonds (sulfohydrolases). The reduced forms are rapidly oxidized by bacteria, forming water and sulfuric acid.

Unlike nitrogen, sulfur is a significant constituent of some rock types. The release of sulfur from mineral weathering generates acidity as sulfur is oxidized.

For example, iron pyrite may weather to form jarosite, releasing 2 H^+ for every three sulfur atoms in pyrite:

$$12FeS_2 + 45O_2 + 30H_2O + 4K^+ \rightarrow 4KFe_3(OH)_6(SO_4)_2 + 16H_2SO_4$$

Sulfur minerals weather rapidly and are not common in well-aerated soils. These reactions are most important when formerly anaerobic layers are exposed to the atmosphere, as occurs with strip mining or draining of flooded areas.

Sulfur is not adsorbed by the CEC, but it may be specifically adsorbed to oxides of iron and aluminum. Sulfate may displace H_2O or OH^- groups from the metal oxides (see chap. 2). If water is displaced, pH is not directly affected but the charge of the sulfate is retained by the complex and the CEC is increased. If one OH^- is displaced, one H^+ will be consumed, and CEC will increase by one negative charge. If two OH^- are displaced, CEC will not change and 2 H^+ will be consumed. Conversely, if adsorbed sulfate is released to the soil solution, acidity may also be released.

If a soil horizon is at equilibrium with the soil solution, no net adsorption or desorption of sulfate occurs. However, disequilibrium may be evident if sulfate concentrations in the solution are depleted (such as by plant uptake) or enriched (such as by acidic deposition). A great deal of current research interest centers on patterns of sulfate adsorption and desorption. These studies typically characterize a soil's ability to remove added sulfate from a solution across a range of concentrations (fig. 6.1). The extent of sulfate adsorption by soil is a function of the quantity added for a given amount of adsorption surface. At low additions, sulfate is strongly retained and this affinity diminishes with greater inputs until

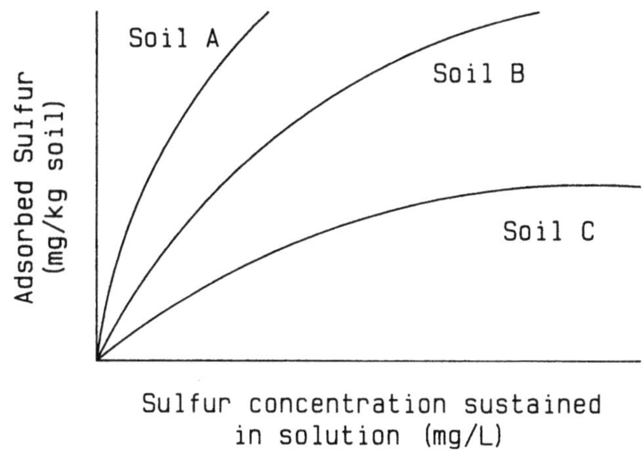

Figure 6.1. Hypothetical sulfate adsorption curves for three soils. Soil A adsorbs most of the added sulfate and maintains a low solution concentration of S. Soil C adsorbs little S and maintains a high concentration of S in solution. Soil B is intermediate. Soil A would show strong resistance to acidification from inputs of sulfuric acid, and soil C would be very sensitive.

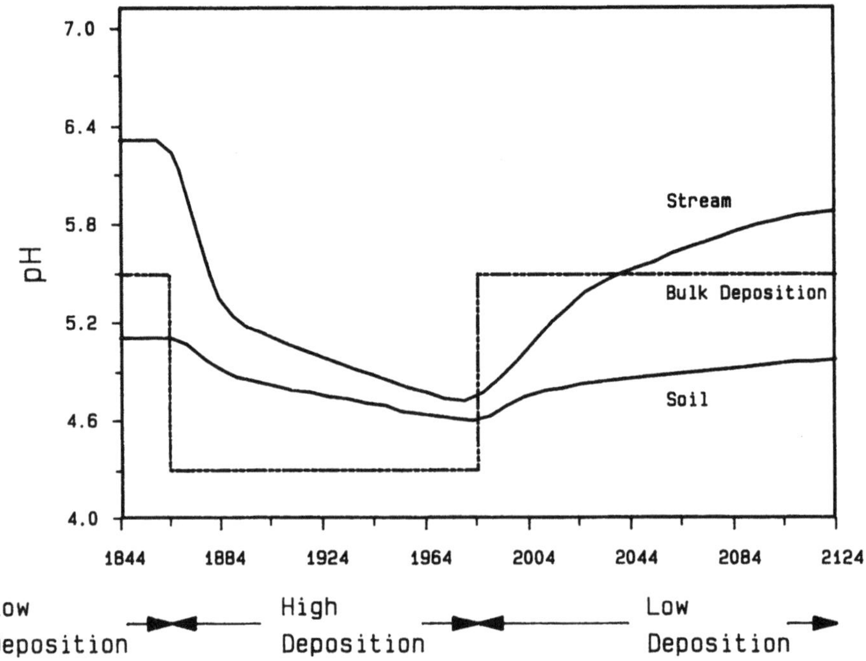

Figure 6.2. Simulations with the MAGIC model show that the pH of soil solution and streamwater gradually decline with the sudden increase in acidic deposition (= decrease in pH of precipitation). Both the stream and soil solution remain above the pH of precipitation. When precipitation pH is reset to earlier levels, the recovery of soil and stream pH is delayed by the gradual desorption of sulfate. (Reprinted with permission from Cosby, B., G. Hornberger, J. Galloway, and R. Wright. 1985. Timescales of catchment acidification. Environmental Science and Technology 19: 1144–1149. Copyright © 1985, American Chemical Society).

the soil is saturated and cannot retain additional sulfate. This process is complicated by the pH-dependence of sulfate adsorption. As the soil acidifies, surface adsorption sites protonate and increase the positive surface charge, enhancing the sulfate adsorption. Conversely, increases in pH causes deprotonation of the exchange surface and decreases both the net positive charge of the surface and the extent of sulfate adsorptions. Some of the removal of sulfate from solution may also be due to biologic immobilization, and microbial activity may account for the majority of the sulfur removal from solution in some soils (David et al. 1984).

Sulfate adsorption appears to be an important mechanism for neutralization of acidic deposition. However, it may be a temporary phenomenon that fails to occur after soil horizons equilibrate with the higher input rates. If adsorption is readily reversible (Singh 1984), desorption could also delay recovery of acidified systems after sulfate deposition was reduced (fig. 6.2). Sulfate adsorption patterns are fundamental to the computer simulations reported in chapter 7.

Very little quantitative information has been developed for sulfur cycles of forest in the South. Switzer and Nelson (1972) estimated that a 20-yr-old stand of loblolly pine required about 0.13 kmol-S/ha (4 kg-S/ha) annually; about 22% of this requirement was met by internal recycling of sulfur from senescent needles. The annual increment in woody biomass was about 0.03 kmol-S/ha (0.8 kg-S/ha). D.W. Johnson et al. (1982) provided the most detailed sulfur budget for a southern forest. They reported that the sulfur requirement of a chestnut oak (Q. prinus L.) forest was about 0.7 kmol-S/ha (22 kg-S/ha) annually, with about 10% of this supplied by recycling from senescent leaves. Annual accumulation of sulfur in biomass was about 0.07 kmol-S/ha (2.1 kg/ha). Given the current deposition rate of sulfur in the South of about 0.15 kmol-S/ha (5 kg/ha) annually (chap. 3), it is clear that deposition exceeds the rate of accumulation in biomass (and perhaps even the rate of uptake) of most forests in the region. Therefore, retention of sulfur deposited from the atmosphere probably depends on geochemical processes rather than on vegetation accumulation (see D.W. Johnson 1984).

About half of the ecosystems listed in table 6.3 appear to be accumulating sulfur (outputs are less than inputs) and about half appear to have net losses of sulfur. The input rates do not include dry deposition, so actual rates of accumulation may be higher than those listed. However, it appears that many forest soils in the South are not capable of retaining current rates of sulfur deposition and that soil acidification and leaching losses of basic cations are potential problems. This conclusion differs somewhat from earlier evaluations. Rochelle et al. (1987) concluded that data from intensively studied sites showed that 20% to 90% of incoming sulfate was retained within ecosystems that were south of the limit of glaciation. Similarly, Rochelle and Church (1987) concluded steamwater data indicated that most incoming sulfur did not pass through to the streams. Our tabulation includes some sites not considered in the earlier studies that fail to accumulate S. Further, some additional retention (or reduction) of sulfate may occur between the bottom of the rooting zone and the location of sampling in aquatic ecosystems. Clearly, additional site-specific data are needed before precise generalizations can be made.

The geochemical processes of adsorption appear critical in determining the retention deposited sulfate, but adsorption isotherms have been determined for only a few soil types in the South. D.W. Johnson et al. (1986) found that sulfate added to two Ultisols in Tennesse was rapidly leached, with no net adsorption in the soil; there actually appeared to be a net release of sulfate previously held within the soil. In this study, phosphate was added along with sulfate, and competition between these anions for adsorption sites may explain part of the pattern.

The Nutrient Cation Cycles

The cycles of nutrient cations (such as calcium, magnesium, and potassium) are simpler than those of nitrogen and sulfur as these elements do not undergo oxidation/reduction reactions in the natural environment. These cycles are intimately connected with ecosystem acidity because negative charges on the CEC, in the soil solution and in plants, must be balanced by H^+ or by these ions. In essence, these cations represent the "dissociated" acids of the ecosystem. These cations are also important because low supply rates can limit forest productivity. The deposition of sulfuric and nitric acids may increase cation leaching from ecosystems simply by increasing the anion concentrations of the soil solution. Some of the anionic charge is balanced by cation nutrients that leach as the water leaves the rooting zone of the forest. D.W. Johnson et al. (1985b) estimate that high values sulfate deposition in Tennessee have increased leaching losses of cation nutrients by two- to threefold.

Nutrient cations are added to available pools within ecosystems through weathering of soil minerals and from atmospheric deposition. Nutrient cations in the atmosphere come from sea salts or terrestrial dust. In both cases, they tend to be accompanied by basic anions (particularly bicarbonate), and deposition of nutrient cations usually entails deposition iof alkalinity.

Mineral weathering usually consumes H^+, but the precise number depends on a series of reactions. For example, the weathering of pyroxene to form kaolinite appears to consume 10 H^+ for every two units of pyroxene weathered:

$$2CaAl_2SiO_6 + 10H^+ \rightarrow Al_2Si_2O_5(OH)_4 + 2Ca^{2+} + 2Al^{3+} + 3H_2O$$

However, the aluminum ions will hydrate and may release a varying number of H^+, depending on soil pH and whether they precipitate as gibbsite (see chap. 2). The net consumption of H^+ is often considered to equal the charge of basic cations released; but at low pH values, the aluminum ions will retain most or all of the H^+ in the waters of hydration, so the actual consumption of H^+ weathering can be higher.

Weathering depends on mineralogy, climate, and the effects of biologic activity. In general, fresh parent material with a high content of basic cations breaks down rapidly as weathering processes exploit imperfections in crystal lattices. Weathering rates probably stabilize after many of the irregularities have weathered and after a rind of weathered material coats unweathered minerals. Finally, weathering rates may apprach zero after the pools of weatherable primary minerals are exhausted (M. Velbel, pers. comm. 1987). Actual rates of weathering are difficult to determine, and very few are available for forest ecosystems (Fölster 1985). Nilsson (1986) tabulated rates from Nordic countries, which ranged from about 0.1 to 1.0 $kmol_c$/ha of H^+ consumpution annually. In the U.S. Pacific northwest, rates have been reported to range from about 1 to 7 $kmol_c$/ha annually (see Kimmins et al. 1985).

In the South, the only available data come from the Southern Appalachians and from Maryland. Velbel (1985) thoroughly characterized soil mineralogy and estimated weathering rates at Coweeta, he concluded that from 0.4 to 1.3 kmol H^+/ha are consumed annually. The rates varied with watershed mineralogy and the extent of secondary mineral formation. In the Catoctin Mountains in Maryland, Katz et al. (1985) estimated a weathering rate of about 1.6 kmol H^+/ha annually in a soil derived from metabasaltic minerals (about 0.3 kmol H^+/ha was due to weathering of calcite). Moving down to the Piedmont of Maryland, Cleaves et al. (1970, 1974) estimated weathering in two watersheds of contrasting mineralogy. A watershed dominated by serpentine rocks (magnesium calcium carbonates) showed a weathering rate of about 3.5 kmol H^+/ha annually. A schist-dominated watershed had a weathering rate of about 0.6 kmol H^+/ha annually.

We know of no studies of weathering in the Piedmont south of Maryland or in the Coastal Plain. Van Lear et al. (1983) assumed that the quantity of nutrient cations leaching from a 41-yr-old loblolly pine forest in South Carolina on a Pacolet series Typic Hapludult could be used as a measure of weathering and estimated an annual rate of only 0.02 $kmol_c$/ha. Although leaching rates are not likely related directly to weathering rates, Ultisols with low contents of weatherable primary minerals probably do have weathering rates approaching 0 (M. Velbelk, pers. comm. 1987). The current Integrated Forest Study (funded by the Electric Power Research Institute through Oak Ridge National Laboratory) will provide estimates for several Piedmont sites (North Carolina, Tennessee, and Georgia) and one Coastal Plain site (Florida).

When nutrient cations are assimilated by vegetation or microbes, an equal amount of H^+ must be released to the soil to maintain charge balance (see chap. 2, and also Binkley and Richter 1987). Therefore, the accumulation of biomass in a productive forest entails acidification of the soil exchange complex. As noted in chapter 2, acidity and alkalinity are conserved, so the increase in soil acidity must be balanced by an increase in plant alkalinity. When biomass decomposes and the nutrient cations are released, acidity is consumed. Therefore, this mechanism of soil acidification is important during stand development, but is reversed when a forest is burned or decomposes. If forest products are removed from the site, however, the acidification effect remains unbalanced.

The amount of cations used to produce new tissues ranges from about 2 to 20 $kmol_c$/ha annually for forests in the South (table 6.4). Most of the cations assimilated are returned to the forest floor each year, and annual accumulation of cations in biomass typically range from less than 1 $kmol_c$/ha to about 3 $kmol_c$/ha.

More than half of the ecosystems in table 6.3 are losing more nutrient cations than they receive in bulk precipitation (if the deposition estimates are accurate). The implications are unclear, however, because rates of weathering may be high enough to compensate for the losses. More estimates of weathering rates for southern soils are needed before complete input/output budgets can be synthesized.

Cycles of Other Elements

Acidic deposition may decrease the availability of phosphate and molybdenum through effects on soil pH and the ionic composition of soil solution, but precise interactions are not well established. The availability of other micronutrients (iron, zinc, manganese, and copper) increase with decreasing soil pH (Lindsay 1979).

Table 6.4. Cation nutrient pools ($kmol_c/ha$) and cycling rates ($kmol_c\ ha^{-1}\ yr^{-1}$) for low-elevation forests in the southeastern United States

Location, Vegetation, Age (in yr)	Biomass Pool	Requirement	Litterfall	Reference
Boone County, MO, Oak species, 35–92	35.6	4.6	2.3	Rochow 1975
Norman, OK, Oak species, 80	82.6	14.0	4.6	Johnson and Risser 1974
Walker Branch, TN, yellow poplar	46.1	6.7	2.9	D.W. Johnson et al. 1985b
Chestnut oak	97.9	6.7	2.5	
Durham County, NC, loblolly pine, 16	17.3	2.9	2.5	Wells and Jorgensen 1975
Gainesville, FL, slash pine, 26	16.1	1.7	1.5	Gholz et al. 1985

Under acidic conditions, elevated concentrations of aluminum may induce precipitation of variscite ($AlPO_4 \cdot 2H_2O$) (Lindsay 1979). Assuming chemical equilibrium with $Al(OH)_3$ and variscite, concentrations of orthophosphate would be reduced by two orders of magnitude following acidification from pH 6 to pH 4. Nevertheless, phosphate minerals exhibit relatively high solubilities and precipitation reactions are probably not critical in regulating solution concentrations in forest soils.

Adsorption of phosphate onto free aluminum and iron oxides is probably the most important mechanism regulating phosphate concentrations in soil solution. These oxide surfaces protonate with decreasing pH, producing positively charged sites in acidic soils. These sites greatly facilitate the retention of phosphate anions. Huang (1975) studied the adsorption of phosphate on particulate aluminum oxide. He observed that adsorption increased with decreasing pH, reaching a maximum at pH 4.5. While Huang (1975) evaluated relatively high concentrations of phosphate (100 to 1000 μmol/L), his observations of phosphate/aluminum interactions may be generally applicable to lower concentrations found in acidic soil solutions. Dickson (1978) observed that addition of phosphate (1.6 and 3.2 μmol/L) to acidic waters containing high aluminum concentrations was followed by phosphate removal from the solution. This retention of phosphate was strongest at pH 5.5. Dickson (1978) suggested that aqueous aluminum may substantially alter phosphorus cycling in acidic waters through adsorption or direct precipitation reactions.

These limited studies do not provide a complete picture of the effects of soil acidification on phosphate availability; more research is needed. The effects of soil acidification on molybdenum availability may follow the same trends as phosphate availability. However, no evidence of molybdenum deficiency has been reported for forest in the South (it is used only in the enzymes that reduce nitrate and fix nitrogen), although deficiencies in agricultural systems do occur in the region (Berger 1962). We do not expect molybdenum availability to be a critical feature of the impacts of acidic deposition, but we would encourage more direct experimentation on molybdenum nutrition of forests.

Nutrient Limitations on Forest Growth in the South

The growth of most forests is limited by the supply rate of nutrients. The degree of limitation and the nutrients responsible for the limitation vary among sites. In the South, most stands are limited by the supply rates of phosphorus and nitrogen. More than 2 million hectares of pine stands have been fertilized in the South with one or both of these nutrients. Less is known about the degree to which cation nutrients limit growth in the South, but evidence is accumulating that potassium deficiencies may be common on some sandy soils and soils with deep, sandy surface horizons (see later discussion). There is no evidence that sulfur supply rates limit forests in the region and, as noted earlier, current rates of atmospheric deposition are likely to insure adequate sulfur supplies for tree growth.

The development of nutrient stresses can include two aspects of nutrient cycles. The first is the nutrient supply rate, a measure of the ability of the soil to supply nutrients to vegetation. The second is nutrient limitation, which is the degree to which the current supply rate prevents the vegetation from achieving maximum growth rates attainable within other environmental constraints (fig. 6.3) (see also Chapin et al. 1986, Binkley 1986). The assessment of nutrient supply rates and limitations in forest is problematic; common approaches have focused on soil fertility measures (to index the nutrient supply rate), on analysis of the nutrient content of foliage (to index nutrient limitation), or on soil classification (based on results of fertilization trials).

In the South, nitrogen limitations have been diagnosed primarily by soil classification or foliar analysis. In the southeast Coastal Plain, researchers at the University of Florida have devised a soil classification scheme that predicts growth responses of slash pine and loblolly pine to phosphorus and nitrogen fertilizers. With the exception of excessively drained soils (Psamments), the probability of growth response to fertilization with nitrogen or N + P in established plantations (table 6.5) is greater than 75% (with response defined as at least 9 to 14 m^3/ha over 5 to 8 yr). Based on foliar analysis, southern pine stands with less than 1.15% nitrogen in current, upper crown needles have a strong probability of responding to nitrogen fertilization. Based on foliar nitrogen data from 62 field trials of the North Carolina State Forest Nutrition Cooperative (NCSFNC), more than two-thirds of established pine plantations fall below this critical level (fig. 6.4). More than half of the same stands had 0.10% P and would probably respond to phosphorus fertilization (fig. 6.5). About 15% of the stands fell below a concentration of K of 0.35% (fig. 6.6).

These widespread responses to nitrogen fertilization indicate that deposition

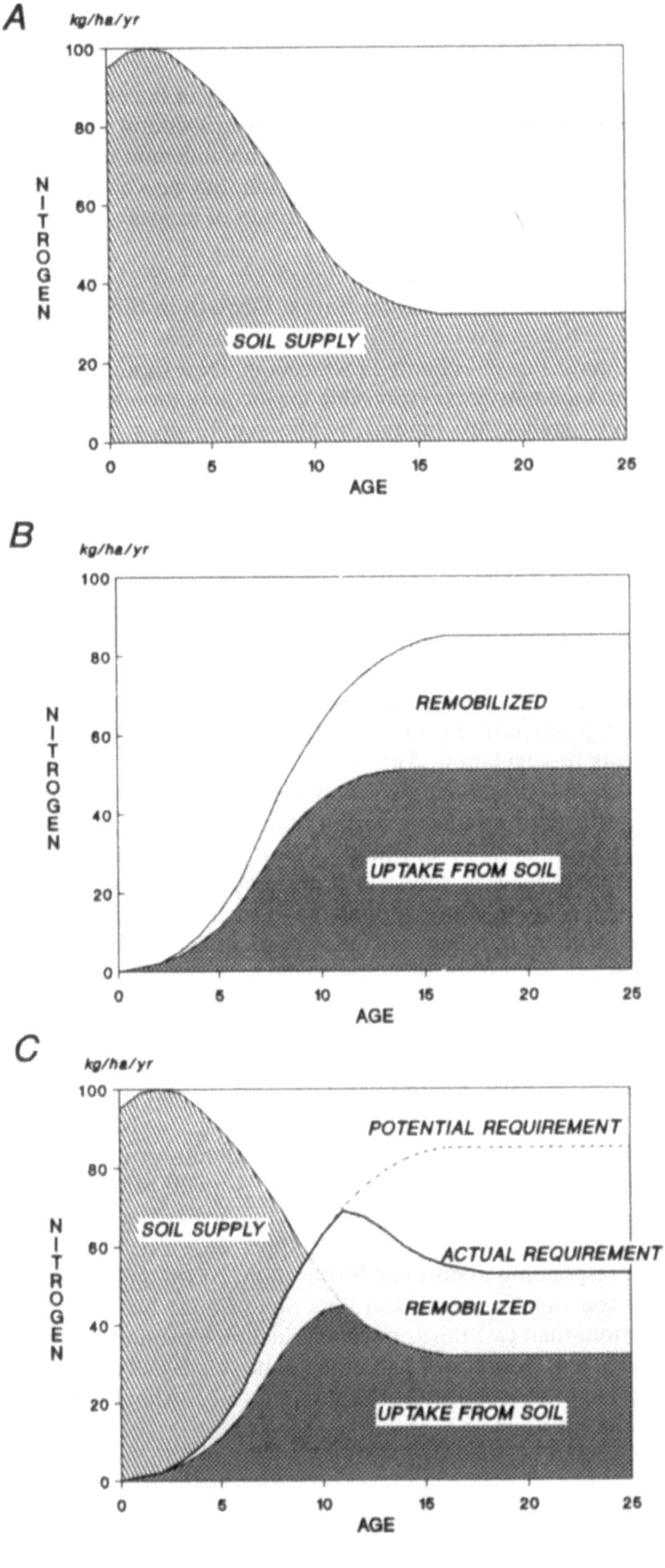
A
kg/ha/yr
NITROGEN
SOIL SUPPLY
AGE
B
kg/ha/yr
NITROGEN
REMOBILIZED
UPTAKE FROM SOIL
AGE
C
kg/ha/yr
NITROGEN
POTENTIAL REQUIREMENT
SOIL SUPPLY
ACTUAL REQUIREMENT
REMOBILIZED
UPTAKE FROM SOIL
AGE

of nitric acid has little opportunity to acidify soils in pine plantations. Uptake by the vegetation neutralizes the acidity and the deposited nitrate may actually increase growth. There may be four exceptions to this generalization. First, sandy soils that are low in organic matter (e.g., Psamments) may have a limited ability to utilize added nitrogen because of water limitations. Furthermore, these soils may have very low capacity to buffer acid inputs. Second, stands on soils that are strongly deficient in phosphorus or potassium may have limited abilities to utilize additional N. Third, ecosystems dominated by nitrogen-fixing black locust may have nitrogen cycling rates that exceed the uptake capacity of the vegetation. Finally, the high elevation forests of the Southern Appalachians appear to have very high mineralization rates (described earlier), and vegetation uptake may not increase to match deposition rates.

Limitations of forest growth by the supply of cation nutrients in the South have been tested in only a few fertilization trials. For example, MacArthur and Davey fertilized a loblolly pine plantation with dolomitic lime, with a calcium/phosphate mixture, and with a combination of both treatments. Lime (applied at 112 $kmol_c$/ha calcium and 107 $kmol_c$/ha magnesium) increased height of the pines at age six by 40%. The concentrations of calcium and magnesium in foliage were relatively high on the control plot (0.17% Ca and 0.11% Mg) and did not increase with liming. However, liming reduced the concentration of aluminum from 270 μg/g to 170 μg/g. The plots treated with a calcium/phosphate fertilizer (19 $kmol_c$ Ca/ha, 6 kmol P/ha) showed a 90% increase in height at age six over the control; nutrient cation concentrations were not increased by fertilization, and in this case aluminum concentrations increased to 300 μg/g. It is difficult to separate causes from correlates in this study. In the liming treatments, increased height might have resulted from the reduction in aluminum in the soil and foliage. However, high aluminum concentrations did not prevent a large increase in growth when phosphorus was also applied.

J. Pye (pers. comm. 1987) fertilized single-tree plots of red spruce, Fraser fir, and Norway spruce with calcium and magnesium at high-elevation sites in the Southern Appalachians. After one growing season, no response in needle size was evident, so no large growth response is expected.

Some preliminary evidence of cation nutrient limitations is available from the University of Florida's Cooperative Research in Forest Fertilization Program (CRIFF), and North Carolina State's Forest Nutrition Cooperative. Comerford et al. (1983) reported that CRIFF soil groups A, B, and C might respond to added potassium (and perhaps to added micronutrients) once growth limitations

---

Figure 6.3. Nitrogen nutrition of stands depends on the supply rate of N from the soil and the potential ability of the stand to utilize the supply (the requirement). The supply may be high early in stand development, and decrease over time (*A*). The stand's potential to utilize N may initially increase with age and then plateau (*B*). Some of the stand's N requirement is met through internal recycling of N. The difference between the supply rate and the potential requirement of the stand represents the degree to which productivity is limited by an insufficient supply (C).

Table 6.5. Classification for soils of the southeastern Coastal Plain with estimated magnitudes and probabilities of fertilizer response for loblolly and slash pine stands

Soil Group	Drainage	Diagnostic Criteria	Taxonomic Subgroup	Response to P at Planting[a]		Response to N or N + P in Established Stands[b]	
				Probability	Volume gain	Probability	Volume gain
				%	m^3/ha/yr	%	m^3/ha/yr
A	Very poorly to somewhat poorly drained	No spodic horizon; argillic within 20 inches	Typic, Albic, Plinthic, and Umbric Aquults	>90	2.8–5.6	>90	2.8–5.6
B	Very poorly to somewhat poorly drained	No spodic horizon; argillic below 20 inches	Arenic and Grossarenic Aquults, Aquents, and Aquepts	75	1.4–4.2	>90	1.4–4.2
C	Very poorly to somewhat poorly drained	Spodic and argillic present	Ultic Aquods and Humods	>75	1.0–2.0	75	1.8–3.5
D	Poorly to moderately well drained	Spodic but no argillic horizon	Typic, Aeric, and Arenic Aquods and Humods	50	0–1.0	75	2.8–7.0
E	Moderately well to well drained	No spodic horizon; argillic within 20 inches	Typic and Plinthic Udults	<50	0–3.5	>75	2.0–4.2
F	Moderately well to well drained	No spodic horizon; argillic below 20 inches	Arenic and Grossarenic Udults, Umbrepts, and Ochrepts	<50	0–3.5	75	0.7–2.8
G	Somewhat excessively to excessively	No spodic horizon; argillic may or may not be present	Psamments	——— Fertilizer ———			

H	Very poorly to poorly drained	Organic surface greater than 20 inches	Medisaprists and Histic Humaquepts	<75	0–4.2	>75	0–4.2

[a]Volume gains represent typical response over a 25-yr rotation. Probability is that for a given stand responding over 15 ft^3/ac/yr.
[b]Volume gains represent typical response over a 5- to 8-yr period in mid rotation stands. Probability is that for a given stand responding over 25 ft^3/ac/yr.
From Allen 1987, used by permission of the Society of American Foresters, modified from Fisher and Garbett 1980, Fisher 1981.

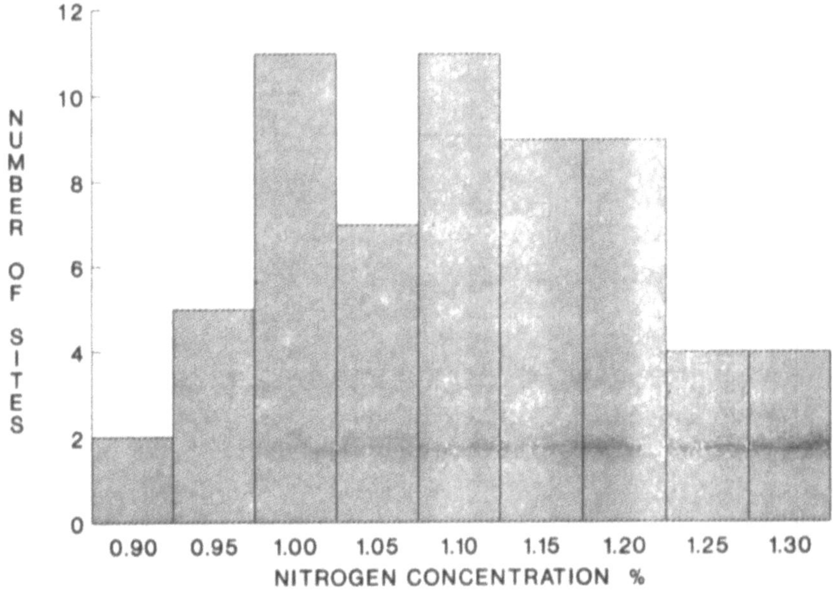

Figure 6.4. Distribution of nitrogen concentrations in first-year foliage of loblolly pine across the South.

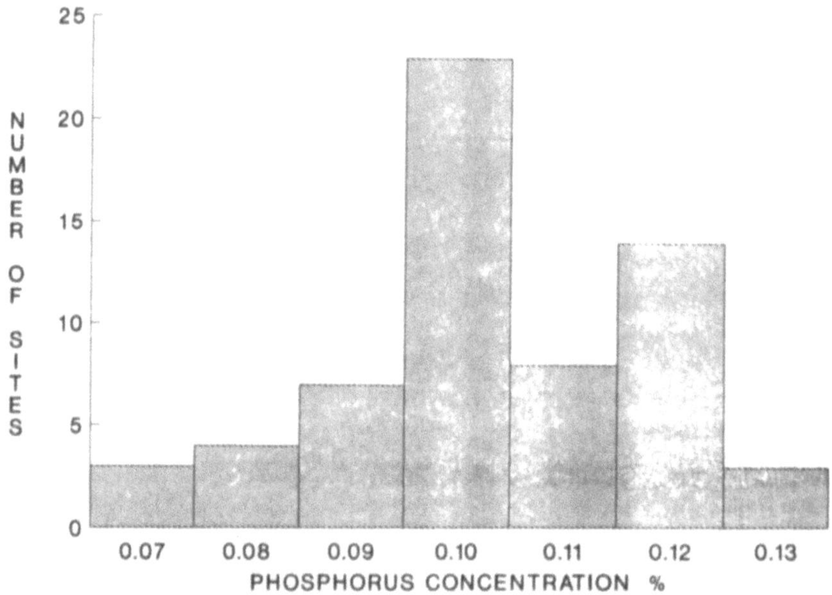

Figure 6.5. Distribution of phosphorus concentrations in first-year foliage of loblolly pine across the South.

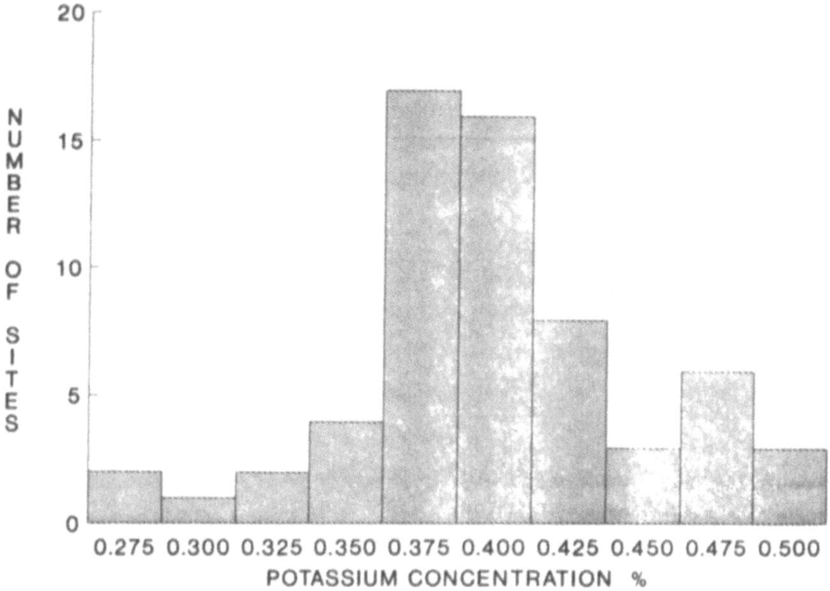

Figure 6.6. Distribution of potassium concentrations in first-year foliage of loblolly pine across the South.

from nitrogen and phosphorus supplies were alleviated by fertilization. They estimated that potassium fertilization might increase growth by an additional 1.1 to 1.4 m^3/ha annually for an unspecified number of years. McFee et al. (1987) found that manganese availability appeared to limit response to NPK fertilization. Five-year analysis of data from the NCSFNC Regionwide no.2 study indicated that pines with less than 0.32% potassium might respond to potassium fertilization with an increase of about 11 m^3/ha extra growth (NCSFFC 1980). At several sites, the apparent response to added potassium continued for at least eight years (NCSFFC 1982). Growth response to fertilization on an upland loam site is illustrated in figure 6.7.

Some additional evidence of potassium limitation comes from productivity studies that relate growth to measured site factors. Turvey and Allen (1987) analyzed the four-year growth responses of loblolly pines in NCSFNC Regionwide no.7 study. They found variations in soil extractable potassium accounted for 60% of the growth differences across the regions, and the apparent effect of potassium supply on growth was greater than the treatment effects (combinations of chopping, piling, windrowing, bedding, subsoiling and disking) that the regionwide study was designed to test. Plots with high concentrations of potassium in soils and foliage showed from 1.0 to 1.6 m greater height after four years than on similar plots with lower potassium supplies. Height growth appeared lower for sites with less than 0.5 $kmol_c$/ha (20 kg/ha) extractable potassium in the top 10 cm of soil than on sites with greater supplies. The strong relationship between potassium and height growth led the authors to conclude

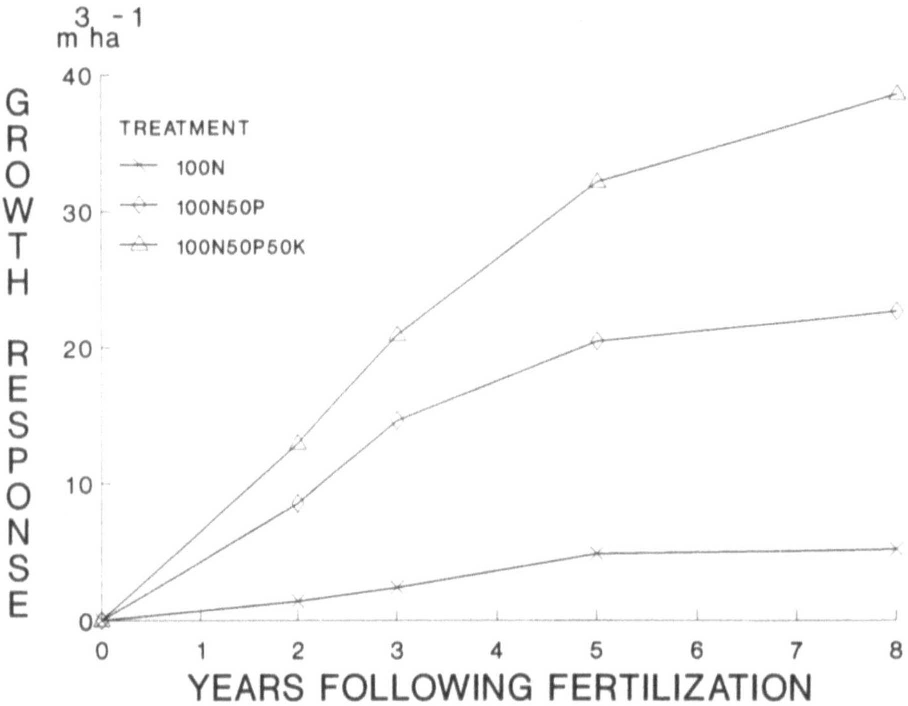

Figure 6.7. Increase in gross volume growth following fertilization in a loblolly pine plantation on an upland loam soil. The addition of potassium fertilizer almost doubled the response to N + P fertilizer.

that potassium limitations of pine productivity are likely but that well-designed fertilization studies are needed to verify the existence and magnitude of response.

Based on preliminary evidence from these and other studies, Allen (1987) estimated critical values of cation nutrients in loblolly pine foliage to be 0.35% K, 0.12% Ca, or 0.07% Mg. Sites with lower concentrations may be expected to respond to changes in the supply rates of these nutrients. About 15% of pine plantations in the South fall below the critical potassium value and about 5% fall below the critical concentrations for Ca and Mg. Such sites might respond to fertilization with growth increases or to accelerated leaching losses with growth decreases.

Conclusions

Forests in the South are unlikely to be limited by the supply of sulfur in the soil, so retention of sulfur deposited by the atmosphere is likely to be controlled by geochemical processes (such as specific anion adsorption) or perhaps incorporation into stable soil organic matter. Few conclusions can be reached on ability of forest soils in the South to retain sulfur. However, ecosystem budgets show that perhaps half of the forest ecosystems in the South currently do not retain sulfur deposited from the atmosphere.

Nitrogen supply rates limit the growth of most forests in the South, so southern soils should retain and neutralize deposited nitric acid. We expect no acidification would occur from nitric acid deposition, with the possible exceptions of: excessively drained, sandy soils; soils with strong phosphorus and potassium limitations (in the absence of fertilization); black locust forests; and high elevation forests. These possible exceptions warrant testing of the ecosystem's capacity to retain added nitrogen.

Although conclusive data are not available, we expect a significant proportion (perhaps 5% to 15%) of pine plantations in the South are limited by the current supply rates of nutrient cations (especially potassium). High leaching rates of nutrient cations might be expected to decrease forest productivity. This expectation warrants testing.

7. Simulation of the Potential Impacts of Acidic Deposition on Forest Soils in the South

Various criteria have been used to evaluate the sensitivity of soils and waters to acidic desposition (Chap. 5), but these criteria remain largely unvalidated. Two approaches to test the relevance of these criteria are computer simulation modeling and long-term field studies. Simulation models have the advantage of rapid evaluation, but the key processes depicted in models need to be tested against field data to develop confidence in model predictions. Long-term field studies can provide definitive evaluations of sensitivity criteria; unfortunately, long-term historical data are currently limited and cannot be used for regional assessments. In the absence of regionally representative data from long-term projects, we used the MAGIC model (Modeling Acidification of Groundwater in Catchments, Cosby et al. 1985a) to evaluate the sensitivity of a representative set of southern forest soils. We chose this model over others currently available because it depicts critical chemical processes that are thought to control soil water acidification and because its limited data requirements allowed us to use information available in the South (unlike more complicated, data-intensive models such as the Integrated Lake-Watershed Acidification Study [ILWAS]) (Goldstein et al. 1985). Sensitivity of soils was gauged by (1) depletion of exchangeable basic cations, (2) increases in soil-solution concentrations of aluminum, and (3) the alkalinity of drainage waters. The results of these simulations (and our sensitivity analyses) cannot provide a final evaluation, but they can be used to test and refine definitions of critical levels for sensitivity parameters (chap. 5). If the simulations show that a substantial portion of soils appear sensitive, more research is warranted (especially on model validation and testing of key parameters). If all soils are well buffered with respect to deposition rates of sulfur and nitrogen, other areas of research on the potential impacts of acidic deposition might be given higher priority.

The MAGIC Model

The MAGIC model was developed to predict long-term reponse of terrestrial and aquatic systems to acidic deposition. To accomplish this objective, Cosby et al. (1985a, 1985b) felt that a process-oriented, physically based model of catchment water was necessary. Therefore MAGIC is a relatively simple, lumped-parameter model that includes processes that are important in regulating streamwater quality over long periods while restricting model complexity and avoiding large data requirements (table 7.1). Spatial variation in biogeochemical processes throughout a soil profile or within a watershed owing to heterogeneity from soil depth or vegetation patterns is not considered. One set of input parameters is used to represent the watershed in simulations aimed at prediction of the chemical composition of streamwater draining the watershed.

Using the approach of Reuss and Johnson (1985, 1986), selectivity coefficients are used to represent cations release through exchange reactions in the soil. Selectivity coefficients are not true thermodynamic constants, but vary markedly from soil to soil as well as with soil conditions, such as pH (McBride and Bloom 1977). Consequently, selectivity coefficients need to be established each time the model is calibrated for a different watershed. These estimates are generally obtained from watershed observations of the distribution of cations on the exchange complex and the cation concentrations in soil solutions and streamwater.

Sulfate adsorption is represented in MAGIC by a Langmuir adsoption isotherm (Cosby et al. 1986a). As with cation exchange, the extent of sulfate adsorption may vary greatly between soils because of differences in soil solutions and soil characteristics (Johnson et al. 1980). The two Langmuir adsorption parameters, the half-saturation constant, and the sulfate adsorption capacity are generally obtained from sulfate adsorption isotherms developed for each soil in batch experiments in a laboratory.

The release of aluminum to solution by the soil is regulated to exchange reactions. However, the distribution of Al on the exchange complex is assumed to be derived from the dissolution of soild phase $Al(OH)_3$. Unlike cation exchange, selectivity coefficients, and Langmuir adsorption parameters, the solutbility of $Al(OH)_3$ is reasonably well defined (Schecher and Driscoll 1987). Users may rely on literature values or solubility data obtained from laboratory experiments.

The remaining thermodynamic constants involving complexation of aluminum with OH^-, F^-, and SO_4^{2-}, the solubility of CO_2 and subsequent dissociation of H_2CO_3 to HCO_3^- and CO_3^{2-} are well defined in the literature (Schecher and Driscoll 1987). These constants can be corrected for variations in temperature with enthalpy values for the reaction of interest as well as for effects of ionic strength (Cosby et al. 1985b).

Rate-dependent processes, such as weathering and element uptake, are not

Table 7.1. Summary of input and output parameters for MAGIC model

Input Parameters		Units
Soil depth		(m)
Porosity		
Soil bulk density		(kg/m^2)
Cation exchange capacity (CEC)		(meq/kg)
Half saturation constant for SO_4 adsorption		(meq/m^3)
Maximum SO_4 adsorption capacity[4]		(meq/kg)
So_4 dry deposition factor		
Al solubility constant for the soil and stream (K_{Al})		
Selectivity coefficients for:		
Al-Ca	(S_{AlCa})	
Al-Mg	(S_{AlMg})	
Al-Na	(S_{AlNa})	
Al-K	(S_{AlK})	
Fraction of exchangeable cation:		
Ca/CEC	(E_{Ca})	
Mg/CEC	(E_{Mg})	
Na/CEC	(E_{Na})	
K/CEC	(E_K)	
% of CO_2 degassing from soil		
Average annual stream flow (Q)		(m/yr)
Average annual precipitation volume (Q_p)		(m/yr)
Background and present precipitation quality		(meq/m^3)
Background weathering rates (W)		(meq/m^2-yr)
Background and present uptake rates		(meq/m^2-yr)
Percentage base saturation (% BS)		

Output Parameters for Soil and Stream for Every Time Step	Units
pH	
ANC	(meq/m^3)
Al^{3+}	(meq/m^3)
Total aqueous Al	(meq/m^3)
Sum of basic cations	(meq/m^3)
Sum of acidic anions	(meq/m^3)
HCO_3^-	(meq/m^3)
Total aqueous F	(meq/m^3)
Total aqueous SO_4	(meq/m^3)

explicitly represented in MAGIC. The release of individual solutes from weathering is constant with time and the rate is simply a function of the H^+ to a fractional order. Similarly, vegetation uptake of nutrients is constant over time. These values are typically obtained through model calibration.

The MAGIC model has been shown to be robust by successfully simulating water quality in the UK (Cosby et al. 1986b, Neal et al. 1986), and the United States (Cosby et al. 1985a, Wright et al. 1986). The model has also been used as part of the EPA's regional assessment of lake sensitivity to acidic deposition (Direct/Delayed Response Program, NAPAP 1986) and is being used with the Watershed Manipulation Project (NAPAP 1986).

Our Approach

For our assessments, we developed a generic set of model parameters (tables 7.2, 7.3) to describe what are assumed to be similar across many sites in the South. We arbitrarily set soil depth at 50 cm, porosity of the soil at 45%, and bulk density at 1.1 kg soil/L. The soil depth of 50 cm approximately corresponds to the rooting zone and therefore is thought to include the most relevant soil horizons controlling acidification effects on forest vegetation. Johnson and Todd (1987) showed that the lower soil/subsoil of a Ultisol in Tennessee was very resistant to acidification owing to sulfate adsorption and mineral weathering. Therefore, our assessments of acidification impacts are not applicable to waters draining deep soils or to surface-water quality in the southeast (see chap. 5), but only to soil solutions in the upper soil horizons. Precipitation averages about 140 cm across the South and runoff averages about 50 cm (chap. 4), and we used these values for the baseline simulations. We varied hydrologic parameters to the actual precipitation and runoff conditions coinciding with the common locations of each soil series (chap. 4). The concentration of CO_2 in the soil atmosphere was set at 0.0315 atm, and soil temperature was 15°C for all soils. Very little information is available on selectivity coefficients for forest soils in the South, so we developed best estimates from the literature (table 7.3) and held these constant across all soils.

The first two exercises involved simulations with four generic soil types: well- and poorly drained clay soils and well- and poorly drained loam soils. These generic types were derived from generalized combinations of the nine representative soils examined in detail later and based on our understanding of forest soils in the southeast (the well-drained soils were given exchange characteristics more representative of Alfisols than of the Ultisols examined in the other simulations). First, the simulations were conducted for each generic soil from 1844 to 1984, and the 1984 chemistry compared with actual data from sites in eastern Tennessee (from Johnson and Todd 1987). These results provided some confidence that the general behavior of the model was consistent with observed soil solutions in the South. Next, simulations were conducted 140 yr into the future to determine which model parameters were most critical in regulating acidification of soil solutions. The third exercise coupled the generic data sets (tables 7.2, 7.3) with specific information on chemistry of the exchange complex for a set of nine soils representing a moisture-by-texture gradient across the South (tables 7.4, 7.5). These data sets were obtained from the regionwide studies of the NCSFNC and provided a uniform set of data that included unbuffered extractions of soil cations. The cation data actually represent 0- to 20-cm depths, but in our simulations, we extended the depths to 50 cm. The estimated chemistry of precipitation in 1844 was used to represent background concentrations of

Table 7.2. Summary of physical input parameters used for MAGIC simulations in the southeast

Reference	Soil Depth	Soil Porosity	Bulk Density	Soil Temperature (C°)	Soil CO_2 (atm)	Streamflow (m/yr)	Precipitation (m/yr)	SO_4 Dry-Dep Factor
Cosby et al. 1985a	1.0	0.45	1.2	—	(0.04–0.01)	0.5	1.1	1.7
Cosby et al. 1985b	(0.3–1.7)	(0.25–0.5)	(0.7–1.3)	—	—	(0.3–0.9)	1.1	(1.4–2.0)
This study	0.5	0.45	1.1	15.0	0.0316	0.5	1.4	1.5

Table 7.3. Summary of soil chemical input parameters for southeastern soils that may be used in the MAGIC model

Data Source	$Log(S_{AlCa})$	$Log(S_{AlMg})$	$Log(S_{AlNa})$	$Log(S_{AlK})$	$Log(K_{Al})$	SO_4-adsorption Half- Saturation (meq/m^3)	SO_4-adsorption Maximum-Capacity (meq/Kg)
Cosby 1985a	4.10	4.10	−0.05	−0.53	9.10	100	8.0
Cosby 1985c	5.80	5.80	0.80	0.32	9.06		
D.W. Johnson and Todd 1987	2.80	3.10	−5.00	−2.90			
D.W. Johnson 1987	4.70	4.50	−2.00	−3.00			
Webb pers. comm. 1987	2.76	3.68	3.31	−0.89		272	1.9
Voice pers. comm. 1987							1.9
This study	4.00	4.20	−0.60	−1.40	9.06	175	2.0

Table 7.4. Matrix of soil series used to represent a range of forest soils in the South

	Drainage class			
Texture	Very Poor	Poor	Somewhat Poor	Well
Clayey	Bayboro	Leaf	Wahee	Pacolet
Fine loamy			Lynchburg	Ruston
Loamy	Pelham			Lucy

Bayboro:	Clayey, mixed, thermic Umbric Paleaquult
Leaf:	Clayey, mixed, thermic Typic Albaquult
Wahee:	Clayey, mixed, thermic Aeric Ochraquult
Pacolet:	Clayey, kaolinitic thermic Typic Hapludult
Lynchburg:	Fine, loamy, thermic Aeric Ochraquult
Ruston:	Fine, loamy, siliceous, thermic Typic Paleudult
Pelham:	Loamy, siliceous, thermic Typic Albaquult
Lucy:	Loamy, siliceous, thermic Arenic Paleudult

ions (table 7.6). Variations in some precipitation chemistry parameters were obtained by interpolation from isopleths in Stensland (1984) to the common locations for each soil series. For the generic soils (and for other precipitation parameters), we used input data from the Electric Power Research Institute/Oak Ridge National Laboratory (EPRI/ORNL) study site at Duke Forest in the Piedmont of North Carolina (K. Knoerr, pers. comm.).

Based on our best estimates of sensitivity criteria from chapter 5, potentially toxic soil solutions were defined as those exceeding 50 μmol Al/L (150 μmol_c/L of Al^{3+}) or where the ratio of Ca/Al dropped below 1 (on a mole basis; 1.5 on a mol charge basis).

Simulation Results

Simulation of future acidification responses are strongly influenced by historical reconstruction of soil chemical processes. The accuracy of MAGIC is in part verified by projecting recent conditions (1984) from a historical reconstruction (starting in 1844) of acidification. In an effort to demonstrate the validity of our modeling effort, we compared simulations with field data from eastern Tennessee (Johnson and Todd 1987). Model predictions for the four generic in soil types showed that simulated, present-day concentrations of major cations and anions were generally comparable to measured data from eastern Tennessee (table 7.7). Simulated concentrations in soil solution were higher for K^+, Na^+, and Cl^- than in the eastern Tennessee sites. This trend was probably due to the use of precipitation data representative of locations closer to the ocean than Tennessee. The simulation results were similar enough to the eastern Tennessee sites to support the assumption that MAGIC provides a reasonable simulation of the dynamics of southern forest soils.

Simulations of future projections (from 1984 to 2124) at current inputs of

Table 7.5. 1984 Values for selected parameters used in model simulations

Parameter	Leaf	Lynchburg	Bayboro	Pacolet	Wahee	Pelham	Lucy	Ruston	Loamy Well Drained	Clayey Well Drained	Clayey Poorly Drained	Loamy Poorly Drained
CEC	54.9	20.7	66.6	31.5	65.7	32.4	15.3	27.0	36.0	33.3	53.1	28.8
E_{Ca}	7.4	4.4	2.0	15.8	2.1	4.1	23.0	11.6	38.4	28.8	4.1	2.5
E_{Mg}	2.3	1.7	1.5	8.6	0.8	1.1	3.4	3.6	4.3	9.4	1.6	1.2
E_{Na}	1.0	1.0	1.0	1.0	1.0	1.0	1.0	1.0	1.0	1.0	1.0	1.0
E_K	0.7	0.7	0.8	2.6	0.5	0.3	0.6	1.3	1.2	1.9	0.6	0.3
% BS	11.4	7.8	5.3	28.0	4.4	6.5	28.0	17.5	44.9	41.1	7.3	5.0
Q_p	1.22	1.24	1.22	1.17	1.17	1.24	1.39	1.57	1.4	1.4	1.4	1.4
Q	0.38	0.36	0.23	0.25	0.20	0.25	0.56	0.76	0.5	0.5	0.5	0.5
W_{Ca}	35.4	32.6	9.7	27.4	11.9	21.1	96.6	83.4	110.9	85.1	41.9	39.7
W_{Mg}	12.6	16.7	11.8	18.4	5.5	6.4	14.8	31.3	12.1	33.0	20.9	28.1
W_{Na}	0.0	0.0	0.0	0.0	0.0	0.0	0.0	2.0	0.0	0.0	0.5	0.1
W_K	0.6	0.3	0.0	0.9	0.0	0.0	0.0	2.2	0.0	0.1	1.2	1.0

[a]See table 7.1 for symbol definitions.

Table 7.6. Assumed background, 1844, and recent, 1984, precipitation concentrations (meq/m^3) use in simulations of acidification of soils in the southeastern United States

Element	Background[a]	Leaf	Lynchburg	Bayboro	Pacolet	Wahee	Pelham	Lucy	Ruston	Generic[b]
Ca[c]	7.0	6.0	6.0	5.0	5.0	5.0	7.0	7.0	6.0	8.0
Mg[b]	4.0	3.0	3.0	3.0	3.0	3.0	3.0	3.0	3.0	3.0
Na[b]	5.0	16.0	16.0	16.0	16.0	16.0	16.0	16.0	16.0	16.0
K[b]	2.0	12.0	12.0	12.0	12.0	12.0	12.0	12.0	12.0	12.0
SO_4[c]	6.0	42.0	42.0	38.0	42.0	38.0	33.0	40.0	33.0	40.0
Cl[c]	7.0	11.0	10.0	11.0	10.0	11.0	14.0	8.0	11.0	30.0
NO_3[c]	8.0	7.0	9.0	5.0	5.0	5.0	6.0	9.0	9.0	32.0
pH	5.47	4.63	4.61	4.74	4.67	4.74	4.81	4.71	4.79	4.2

[a] Cosby et al. 1985a.
[b] Binkley, Duke Forest, pers. comm. 1987
[c] NADP/NTN 1987.

Table 7.7. Recent weighted average annual soil-solution concentrations (μmol (c)/L) and standard deviations

	Loblolly Pine	Mixed Oak	Clayey Well Drained	Clayey Poorly Drained	Loamy Well Drained	Loamy Poorly Drained
pH	5.4	4.9	5.4	5.1	5.5	5.0
Ca^{2+}	207 ± 50	102 ± 23	249	157	315	131
Mg^{2+}	63 ± 12	90 ± 27	95	70	42	79
K^+	27 ± 18	25 ± 11	45	49	45	49
Na^+	23 ± 6	14 ± 4	31	38	30	41
SO_4^{2-}	251 ± 63	228 ± 56	170	170	170	170
NO_3	1 ± 2	0.2 ± 0.4	9	9	9	9
Cl^-	41 ± 22	30 ± 11	84	84	84	84

From Johnson and Todd, 1987, for loblolly pine, and mixed oak sites in eastern Tennessee compared with model simulations for four generic soil types.

Table 7.8. Present-day, 1984, and future, 2124, simulations of base saturation (% BS), acid neutralizing capacity (ANC) and Al concentrations (μmol(c)/L) for four generic soil types

Soil Type	Present % BS	Future % BS	Present ANC	Future ANC	Present Al	Future Al
Clayey Well Drained	41.7	23.6	157.1	110.0	2.5	3.9
Clayey Poorly Drained	7.8	1.8	50.5	−24.1	9.2	42.3
Loamy Well Drained	46.7	28.9	167.9	123.5	2.2	3.4
Loamy Poorly Drained	5.5	2.0	35.0	−15.5	12.5	36.3

acidic deposition on the four generic soils show major changes in base saturation, acid neutralizing capacity, and soil-solution aluminum concentrations (table 7.8). The patterns can be attributed largely to the fact that weathering rates (ranging from 0.6 to 1.2 kmol H^- consumed/ha annually for the generic soils, table 7.5) are too low to balance H^+ inputs (or to balance basic cation leaching). In general, the poorly drained soils (currently showing low base saturation) are most sensitive to acidification over the 140-yr simulations. These soils show declines of ANC to negative values, which indicates that drainage waters would not only fail to supply ANC to streams and lakes, but would also actually contribute acidity and consume ANC supplied from other sources.

Even with the decline in ANC, the poorly drained soils do not approach the critical value we established for solution aluminum concentrations (50 μmol/L, 150 μmol_c/L). Well-drained soils, on the other hand, are well buffered by the exchangeable bases. The simulated decrease in base saturation of the well-drained soils are not sufficient to produce an ecologically important decrease in ANC or an increase in soil-solution aluminum.

The poorly drained clay soil is the most sensitive of the generic soils owing largely to the low CEC and exchangeable basic cations. Therefore, we focused our sensitivity analyses on the generic poorly drained clay soil to evaluate which parameters have the greatest influence on acidification. The most critical variables appear to be sulfate adsorption, soil depth, CEC, and cation selectivity coefficients. When sulfate adsorption capacity is increased from values typical in the upper rooting zone (maximum capacity of 2 $mmol_c$/kg) to a level more typical of lower soil horizons (maximum capacity of 8 $mmol_c$/kg) in southern forest soils, acidification effects are delayed by about 20 yr (fig. 7.1). Essentially no effect is evident when sulfate adsorption capacity is decreased to 0.5 $mmol_c$/kg. Thus, an increase in sulfate adsorption capacity might delay the acidification of drainage water over the short term, but it would appear to have little or no effect beyond the delay period.

Simulations, however, are more sensitive to changes in soil depth, as an increase from 0.5 to 1.0 m greatly restricts acidification of the soil solution (fig. 7.1). This characteristic is evident because a doubling of soil depth also doubles the amount of soil-mitigating acidification through sulfate adsorption and exchange of basic cations. A decrease in soil depth to 25 cm increases the rate of acidification.

When the cation exchange capacity is increased from 53 $mmol_c$/kg to 142 $mmol_c$/kg, the rate of decrease in base saturation is reduced (fig. 7.2). A decrease in soil CEC to 30 $mmol_c$/kg, however, causes a dramatic increase in the rate of acidification. The soil reaches a new, quasi-steady state in 40 yr.

An order of magnitude increase or decrease in the selectivity coefficients (S) has little influence on the changes in base saturation (fig 7.2). However, a decrease in the coefficients increases the aluminum concentration of the soil solution by more than 50%, whereas an increase in selectivity coefficients decreases aluminum concentrations by more than 50%. This pattern indicates that predictions of the ionic composition of soil solutions depends strongly on these coefficients as well as on the base saturation of the exchange complex.

We also examined the acidification effects owing to changes in the input of acidic deposition. Doubling the deposition rate results in a decrease in base saturation at year 2124 from 1.8% (current rate of deposition) to 1.2% (fig. 7.3). Although this change is slight, conditions of low base saturation result in marked sensitivity of soil-solution concentrations to acid inputs. For this reason, concentrations of aluminum in the soil solution during year 2124 increase from 42 μmol_c/L with current deposition rates to 172 μmol_c/L with the doubled rate of deposition. Conversely, a 50% reduction in deposition rate results in a slight improvement in soil-solution quality after 100 yr of simulation (see also fig. 6.2). Model projections suggest that soil base saturation would begin to increase

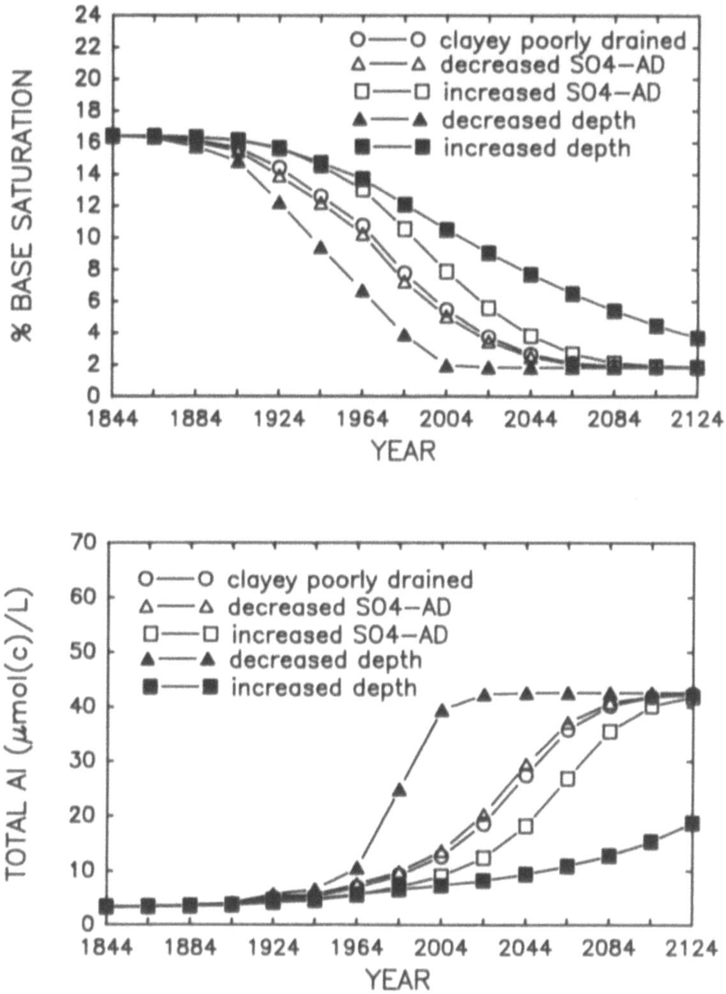

Figure 7.1. Sensitivity analysis for the generic clayey, poorly drained soil to changes in sulfate adsorption capacity and soil depth. Both base saturation (*top*) and the aluminum concentration (*bottom*) of soil solution are much more sensitive to soil depth available for buffering than to the sulfate adsorption capacity.

around year 2074, with a corresponding decrease in soil-solution aluminum concentrations. This simulation did not reach a steady state at the conclusion of the run in 2124, but the trend for improved quality of soil solution was clear.

The final series of simulations used actual exchange complex chemistry for forest soils from across the South. All soils show a trend toward acidification (table 7.9). Acidification of some soils is slight (such as Lucy and Ruston series), whereas others show major change (such as Bayboro and Wahee). Interestingly, the changes in pH range only from 0.0 to 0.3 units, but changes in base saturation, ANC, and solution aluminum concentrations are large and highly variable

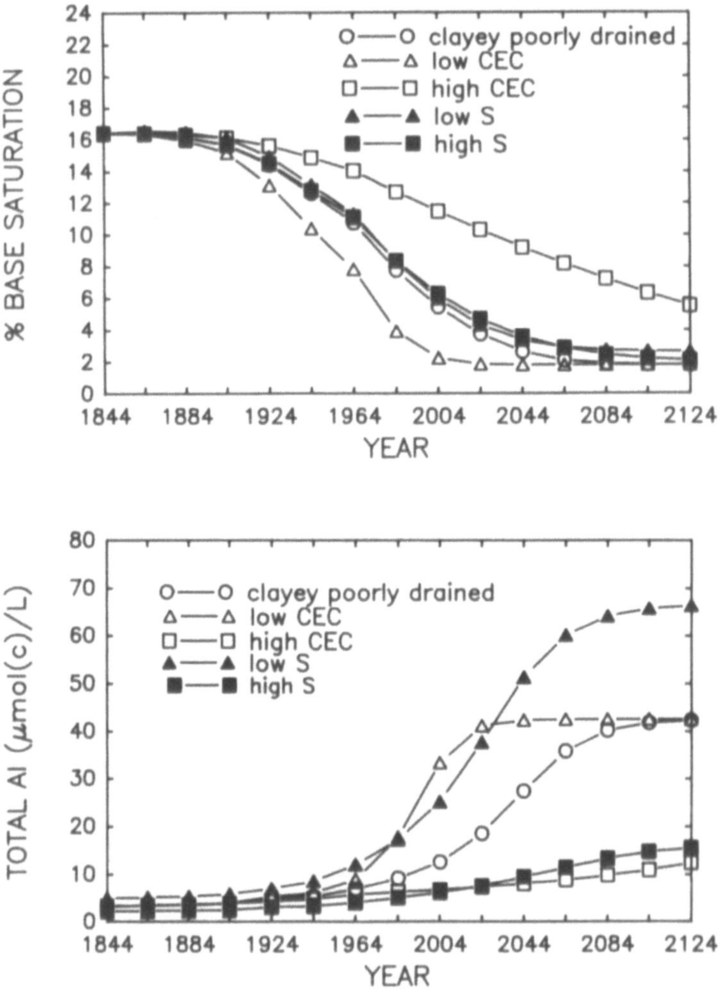

Figure 7.2. Sensitivity analysis for the generic clayey, poorly drained soil to changes in CEC and cation selectivity coefficients (S). Base saturation (*top*) is relatively insensitive to changes in the selectivity coefficients, but the solution concentrations of aluminum (*bottom*) depend strongly on the selectivity coefficients.

across soils. The most sensitive soils are characterized by current values of low base saturation ($<10\%$, based on sum-of-cations CEC), low soil solution pH (<5.0), and low soil-solution ANC (<50 $\mu mol_c/L$). Moderately senstive soils have present-day base saturation between 10% to 20%, solution pH between 5.0 and 5.5, and solution ANC between 50 and 100 $\mu mol_c/L$. Insensitive soils have higher values for each of these parameters.

Soil-solution concentrations of aluminum do not increase significantly until the base saturation decreases below 2%. Uncertainty in the selectivity coefficients could increase this critical level to 5% or more. This pattern also corre-

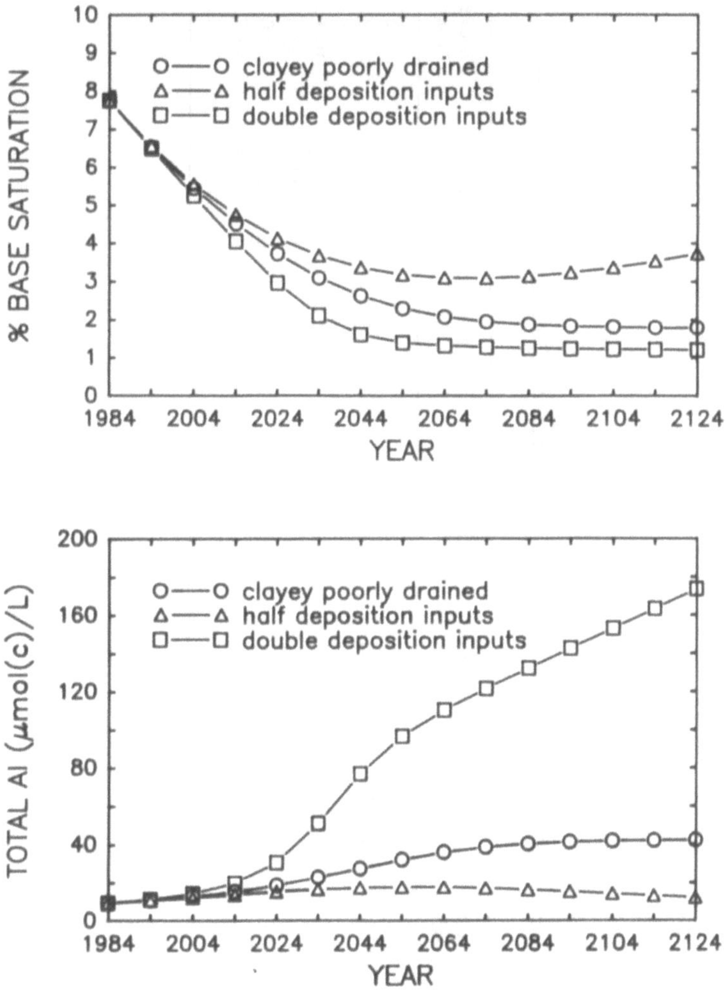

Figure 7.3. Sensitivity analysis for the generic clayey, poorly drained soil to changes in atmospheric deposition rates. Base saturation (*top*) continues to decline even if deposition rates are cut in half, but the concentration of aluminum (*bottom*) in soil solution declines by over 60%. Doubling the deposition rate stabilizes base saturation at only slightly less than current deposition rates, but the concentration of aluminum in soil solution is more than four times higher with the increased deposition.

sponds with ratios of Ca/Al of less than 1 (on an equivalent basis), indicating potential aluminum toxicity problems. A rapid increase in solution concentrations of aluminum was also predicted by Reuss and Johnson (1986) when base-saturation values declined below about 10% (fig. 7.4). Note that the critical level of base saturation (where aluminum concentrations in soil solution increase rapidly) increases with increasing ionic strength (= sulfated added in the simulation) of the soil solution.

Table 7.9. Recent, 1984, and future, 2124, simulations of soil and soil-solution concentrations (in μmol(c)/L) for typical soil serious of the southeastern United States

Soil Series	Present pH	Future pH	Present % BS	Future % BS	Present ANC	Future ANC	Present Al	Future Al	Future Ca/Al
Leaf	5.2	5.0	12.3	4.5	75.6	28.4	6.4	14.7	8.5
Lynchburg	5.1	4.8	8.8	2.9	56.4	3.8	8.8	25.2	4.4
Bayboro	4.9	4.6	6.1	1.7	24.7	−78.4	17.4	86.9	0.8
Pacolet	5.3	5.1	29.3	12.9	116.0	64.6	4.7	8.6	19.1
Wahee	4.9	4.6	5.5	1.7	13.0	−103.8	22.3	109.5	0.8
Pelham	5.0	4.7	8.4	2.6	44.8	−21.3	11.8	41.3	2.9
Lucy	5.4	5.3	30.1	22.1	142.7	120.1	2.7	3.4	56.2
Ruston	5.3	5.3	17.9	13.8	116.3	101.4	3.3	3.9	31.5

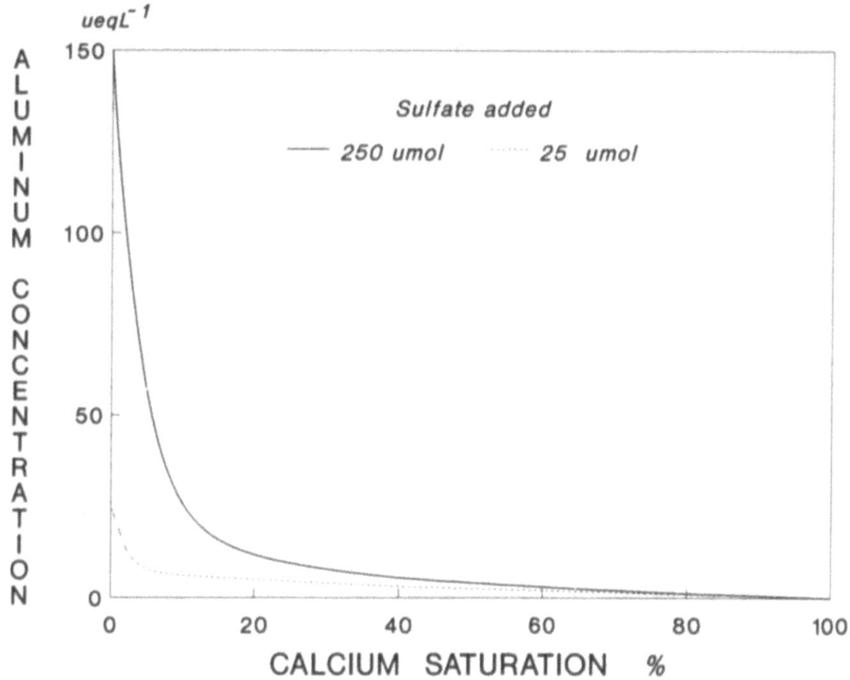

Figure 7.4. Response of soil-solution aluminum concentrations to changes in base (calcium) saturation of the exchange complex, based on the model of Reuss and Johnson 1986.

Discussion of Simulation Results and Implications

The interpretation of simulation results depends on the accurate depiction of processes used in the model as well as the quality of the input data. The version of the MAGIC model we used seems to be well suited for assessments of future acidification of soil solutions and in particular the potential impacts on vegetation because of elevated concentrations of aluminum. However, the model currently has limited usefulness in predicting nutrient deficiencies (such as K^+ limitation of pine growth). Fortunately, empirical fertilization trials can help fill gaps in knowledge relative to nutrient deficiency without decades of data. Furthermore, the MAGIC model does not depict nitrogen cycling, therefore it is not appropriate for addressing questions regarding the effects of elevated nitrogen deposition. Currently, most forests in the South retain nitrogen efficiently; this limitation in our simulations may become more critical in the future if nitrogen availability increases owing to atmospheric deposition or fertilization. A historical reconstruction period is also required by the MAGIC model; since virtually no data are available on soil and vegetation characteristics from such periods, this portion of the modeling effort could potentially result in imperfect predictions of soil and solution chemistry. Therefore, interpretation of MAGIC simulations should focus on trends in chemistry rather than on absolute values.

The mathematical approach used in MAGIC appears sound because the model has simulated watershed stream chemistry successfully throughout the world (Cosby et al. 1985, Cosby et al. 1876, Neal et al. 1986, Wright et al. 1986). This robust behavior reinforces our confidence in the model's performance for assessing regional acidification in southern forest soils. The close agreement between the historical reconstructions of the four generic soils and the actual chemistry of the two sites in eastern Tennessee also provides confidence in the model's performance. Further verification of the model would involve comparisons of simulations with actual long-term changes in soil chemistry (such as the preliminary test with the 20-yr sequence of soil change in South Carolina in chap. 2). Unfortunatley, few long-term field studies are available. Moreover, the simulations indicated that 50 to 100 yr of record might be needed for a thorough test of the model's predictions.

Even though limitations are inherent in any modeling exercise, simulations can lend insight into the more important mechanisms controlling soil-solution chemistry as well as general trends of future conditions. For example, understanding may be obtained on the interaction acidic deposition and tree nutrition by adapting current models to include more realistic nutrient cycling (e.g., availability of phosphorus and cation nutrients). Simulation modeling may also provide information about specific data needs for answering important environmental quality questions.

Previous regional assessments have relied on sensitivity criteria of CEC, base saturation, and soil pH (table 7.10, see also chap. 5). These assessments are generally based on laboratory experiments or best guesses. Model simulations have allowed us to incorporate properties of the soil solution (ANC and Ca/Al ratios) in our suggested criteria. Interestingly, our suggested criteria based on computer simulations are similar to recommendations developed in earlier assessments. Based on our simulations, we expect sensitive soils to demonstrate large changes in base saturation, solution ANC, and aluminum if: CEC (by sum of cations) <100 $mmol_c$/kg, base saturation < 10%, soil solution pH <5.0, and ANC < 50 μmol_c/L. The Bayboro and Wahee soils fall within these ranges. Moderately sensitive soils are characterized by CEC < 150 $mmol_c$/kg, base saturation between 10% and 20%, soil solution pH between 5 and 5.5, and ANC between 50 and 100 $mmol_c$/L. The Leaf, Lynchburg, and Pelham soils fit these criteria. Insensitive soils would have values greater than the range for moderate soils, and include Pacolet, Lucy, and Ruston soils.

In comparison, previous studies have used CEC values from 60 to 150 as the critical level for sensitive soils (table 7.10). These earlier values, however, were based on CEC measured in buffered extractants as pH 7 and comparable sum-of-cations CEC on these soils would be lower (by 20% to 80%). Earlier studies also tended to conclude that soils with moderate pH (> 5) would be more sensitive than those with lower pH. Note that our pH values represent the soil solution, whereas other studies have focused on the pH of soil samples in laboratories that may be 0.5 to 1.0 units lower. Although changes in pH may be more

Table 7.10. Sensitivity criteria from this study and those discussed in chapter 5

Reference	Class	CEC (mmol$_c$/kg)[a]	Base Saturation (%)	pH[b]
McFee 1980	Sensitive	< 60		
	Slightly sensitive	60–150		
	Nonsensitive	> 150		
Klopatek et al. 1980	Sensitive	< 120	30–50	> 5.0
	Slightly sensitive	120–200	30–40	> 5.0
	Insensitive	> 120	< 30	< 5.0
	Very insensitive	> 200	> 50	> 6.0
Olson et al. 1982	Highly sensitive	< 62		> 5.5
	Moderately sensitive	62–90		> 5.5
	Insensitive	> 90		—
Turner et al. 1986	Sensitive:			
	basic cations	< 150	20–60	> 4.50
	Al toxicity	< 150	< 20	< 4.50
This study	Sensitive	< 100	< 10	< 5.00
	Moderately sensitive	100–150	10–20	< 5.50
	Insensitive	> 150	> 20	≥ 5.50

[a] The CEC values for this study are based on unbuffered extracts; the others are based on extracts buffered at pH 7.
[b] Note pH for this study is pH of soil solution; the others are for soil samples in the laboratory.

pronounced above pH 5, our recommended criteria emphasize the sensitivity of soils below pH 5 because our definition is based on the potential for aluminum toxicity or deficiencies in basic cations rather than on rates of change *per se*.

The most notable difference between our recommended criteria and those from earlier assessments is our low value for base saturation. In our simulations, soils with base saturation above 5% to 10% had low concentrations of aluminum in soil solution as well as high Ca/Al ratios. This value varies depending on the choice of selectivity coefficients. A level of 10% base saturation for a soil where CEC was derived from the sum of cations would be equal to an even lower base saturation with CEC measured at pH 7. Therefore, our 10% base-saturation value differs substantially from the 20% or higher values suggested in earlier studies. We feel more work is needed to establish the conditions under which rapid changes in solution concentrations of aluminum occur with decreases in base saturation. For example, the South Carolina soil described earlier in chapter 2 started with a base saturation (by sum of cations) of about 40%, yet it declined to under 10% in the upper horizons in just 20 yr. This soil was very poorly buffered; not all soils would change so rapidly. However, this case study does illustrate marked changes in base saturation may indeed occur and a better quantitative understanding of cation leaching is necessary to refine sensitivity criteria.

Despite the differences in criteria for critical levels of base saturation, our system roughly matched that of Turner et al. (1986) for the aluminum-toxicity

Table 7.11. A comparison of proposed sensitivity criteria for assessing future acidification effects of soil solutions with sensitivity criteria[a] and suggested criteria by Turner et al.[b]

Soil Series	Simulated Conditions 2124			Proposed Criteria			Turner suggested Criteria	
	Al > 50 μmol(c)/L	Ca/Al < 1	ANC < 50 μmol(c)/L	Sensitive	Moderately Sensitive	Insensitive	Nutrient	Toxicity
Leaf			x		x			x
Lynchburg			x		x			x
Bayboro	x	x	x	x				x
Pacolet						x	x	
Wahee	x	x	x	x				x
Pelham			x		x			x
Lucy						x	x	
Ruston						x		x

[a] Defined in table 7.10.
[b] Turner et al. 1986.

hypothesis (table 7.11). By our criteria, five of the eight soils series we examined would be sensitive or moderately sensitive. By the criteria of Turner et al. (1986), the same five soils would be sensitive, along with the Ruston series, which was insensitive by our criteria. Our classification based on basic-cation deficiency did not overlap with that of Turner et al. (1986) for these nine soils. Neither of the two soils identified as sensitive to basic-cation deficiency of the criteria of Turner et al. (1986) were considered sensitive by our criteria. By our simulations, the critical level of base saturation may be much lower than the values suggested by Turner et al., and the conditions become more critical at lower pH values rather than at moderate pH levels. This difference results from our perspective on insufficient supplies of basic-cation nutrients for tree growth contrasted with the focus on the rate of basic-cation leaching by Turner et al. (1986, p. 30).

Conclusions from Simulations

Although simulations of soil acidification cannot provide firm conclusions about the actual sensitivity of soils, they can provide estimates of (1) which parameters appear most important in regulating acidification, (2) whether acidificaticn is likely in any soil type, and (3) which soil types are sensitive. Our simulations with the MAGIC model lead us to several conclusions:

1. Decreases in the pH of soil solutions of just 0.3 units can be accompanied by major decreases in ANC and increases in concentrations of aluminum in soil solution. Based on information discussed in chapter 5, a decrease in pH of 0.3 might be expected for many soil types on a time period of decades; such a decrease may result from acidic deposition or from ecosystem nutrient cycling.
2. Increasing the maximum sulfate adsorption capacity in the simulations by fourfold delayed soil acidification by about 20 yr, but did not alter long-term trends. Coupled with the observation that many forest ecosystems in the South do not currently accumulate sulfur inputs (chap. 6), we expect that sulfate adsorption will be important for determining rates of soil acidification in only a portion of forest soils in the South and even then only for a limited period. The Direct/Delayed Response Program of the EPA focuses heavily on sulfate retention in soils as a major factor in acidification of waters. Our simulations indicate sulfate retention may not be so important when considering acidification effects in the upper soil (rooting zone). However, deep soils rich in sesquioxides may retain large quantities of sulfate, greatly mitigating the acidification of downstream aquatic ecosystems.
3. Decreases in acidic deposition led to at least partial recovery of the most sensitive generic soil. This illustrates the critical balance between deposition rate and the ability of mineral weathering to neutralize inputs. If deposition rates are reduced to below weathering rates, soils should recover by increasing the base saturation of the exchange complex. This balance between rates

of deposition and weathering was emphasized in a recent Nordic workshop (Nilsson 1986).

4. Coupling our simulations with previous assessments, we do not think it safe to assume that a substantial portion of forest soils in the South will not acidify over the next century to the point where the availability of nutrient cations or the toxicity of soil-solution aluminum will reduce tree growth. We conclude that well-focused research on key processes and regional patterns is needed to provide a foundation for drawing stronger conclusions.

The critical areas warranting research are:

1. Better information is needed on the factors that regulate sulfate adsorption, such as the pH-dependent adsorption and competition between anions (sulfate vs. organic anions) and the effects of pH on adsorbing surfaces. This information would allow more accurate depicitions of this process in simulation models. In addition, regional information is needed on the extent of sulfate adsorption to allow more realistic assessments across the South. In the longer run, however, we do not expect this mechanism to prevent acidification.
2. The characteristics of soil exchange complexes (CEC, base saturation, etc.), are much easier to measure on a large number of sites than are soil-solution parameters. However, soil-solution parameters are more critical for evaluating plant responses. Research is needed on the adequacy of simulations of soil-solution chemistry from easily measured properties of the soil exchange complex (including cation selectivity coefficients).
3. The current extent of cation nutrient limitations on forest growth need to be rigorously determined, both within sites and across the region. Critical ranges for foliage and soil chemistry need to be determined.
4. Simulation models and field studies can predict the concentrations of aluminum in soil solution with reasonable accuracy, but the critical levels relative to vegetation need to be determined in field studies by manipulating aluminum concentrations. Southern pines seem relatively insensitive to concentrations of aluminum that are likely to occur in southern soils, but hardwoods may be more sensitive (chap. 5).
5. Long-term monitoring sites need to be established to validate model predictions, and to account for the importance of factors not currently included in models (such as changes in the size of the soil-acidity pool, and changes in the acid strength of the pool).
6. The usefulness of models would be improved by incorporating routines to simulate dynamics of nitrogen as well as the availability of cation nutrients to vegetation. The nitrogen cycling components of the model of Pastor and Post (1986) would be suited to this task, but no models of forest nutrient cycling currently incorporate basic-cation availability routines.

8. Synthesis and Recommendations

A variety of studies from Europe and North America have shown that soil chemistry is fairly dynamic, with substantial changes occurring over a period of decades. The causes behind such dynamics are less clear. Acidic precipitation certainly plays a role at some locations, but ecosystem dynamics are also likely to be important. Forest growth is sensitive to soil chemistry, as evidenced by response to fertilization. To date, forests in the South have responded most strongly to additions of phosphorus and nitrogen, but some evidence of limitation by other nutrients is available. Our conclusions about possible effects of acidic deposition are applicable to more general issues of forest nutrient cycling and productivity, and our suggested research directions would have value regardless of the degree of impact of acidic deposition. The recommended monitoring research, in particular, is needed for a wide range of forest health and productivity concerns.

Rates of acidic deposition in the South are on the order of 0.5 to 1.5 kmol H^+/ha annually. Sulfuric acid comprises about two-thirds of this amount and nitric acid accounts for most of the rest. Most forests in the South are nitrogen limited, so deposited nitrate is effectively retained and the associated acidity is neutralized. Forests in the region have ample supplies of sulfur, and retention of deposited sulfate depends on microbial and geochemical processes. Information on sulfur retention is limited, but input/output budgets show that perhaps half of the forests in the South lose as much (or more) sulfate in drainage water as they receive from the atmosphere. Sulfate leaching can be important even if some net retention occurs.

Forests generate H^+ as a natural part of nutrient cycling, and these processes (such as bicarbonate formation, biomass accumulation of cations) typically match or exceed current rates of deposition. Further, assimilation of the anions of sulfuric and nitric acids neutralizes the associated acidity. Therefore, the potential impacts of acidic deposition need to be evaluated as additions to the background production and consumption of H^+ ions.

Leaching losses of basic cations may have increased by about 0.5 to 1.0 $kmol_c$/ha annually as a result of acidic deposition, perhaps representing a doubling or tripling of background rates (Johnson et al. 1985b). The potential depletion of basic cations is important for several reasons. They are essential plant nutrients, and perhaps 10% to 15% of the commercial pine forests in the South may be currently limited by the availability of these nutrients (especially if any nitrogen and phosphorus limitations are removed by adequate fertilization). The degree of base saturation of the CEC regulates the pH of the soil solution as well as the equilibrium concentration of ions in solutions. As base saturation declines, concentrations of potentially toxic aluminum ions increase. The potential of weathering reactions to resupply the additional cations lost in drainage waters is largely unknown in the South, as is the potential for nutrient uptake from the subsoil. Many soils in the region are old and depleted of readily weathered primary minerals, so we conclude that in the absence of better data, it is not safe to assume that weathering rates are sufficient to prevent depletion of exchangeable cations.

Soil chemistry should be expected to be fairly dynamic on a time scale of decades, based on the few studies available and on our computer simulations. Decreases in pH of 0.5 to 1.0 units over several decades should not be uncommon in some soil types, and such decreases may be associated with very large changes in the chemistry of the exchange complex and soil solutions. However, the implications of such changes are not clear; more information is needed on the availability of nutrient cations to trees and on the sensitivity of trees to aluminum in soil solution. Pines may be relatively insensitive to changes in soil chemistry, but spruces and hardwoods may be affected. Field trials are needed (see later suggestions).

Soil pH (and chemistry of the exchange complex and solutions) may also change because of changes in the quantity of organic matter in soils and because of changes in the acid strength of the exchange complex. Almost no information is available on rates of change in these parameters, but their importance may be of the same order of magnitude as leaching of basic cations.

The state-of-knowledge of sulfur dynamics does not allow adequate prediction of the degree of S retention by forest soils in the South. However, many soils in the region appear to be sulfur saturated (outputs match or exceed inputs), and our simulations suggest that sulfur retention by geochemical adsorption may delay acidification of otherwise sensitive soils, but the delay may be only a matter of one to several decades.

The ability of forests in the South to retain nitrogen may decline after decades of deposition of nitric acid. However, management of commercial forests (including harvesting and burning) will probably prevent nitrogen saturation, and we expect most forests will retain essentially all deposited nitrogen for many decades. Some potentially important exceptions include: forests on very poor soils where limitations of other resources prevent utilization of deposited nitrogen, high-elevation spruce and fir forests where nitrogen availability may already exceed stand requirements, and forests where black locust has accelerated nitrogen cycling to high rates.

Our computer simulations provide a first approximation of the possible impacts of deposition of sulfur and nitrogen compounds on forest soil chemistry; however, we expect that improvements in model parameters (such as cation selectivity coefficients and the incorporation of spatial variability) over the next decade may produce improved simulations that lead to conclusions that differ from ours. However, our simulations suggest that soil chemistry is responsive to rates of atmospheric deposition and that it is not safe to assume that forest soils in the South will not be acidified by acidic deposition. Many soils in the South already have low pH, and they may show only small changes in pH over a period of decades. However, such soils may still be acidified by definitions that are important to tree responses (such as the availability of basic cations and increased concentrations of potentially toxic aluminum) without major changes in pH.

The state-of-knowledge on aluminum dynamics in soil solution and effects on vegetation is very incomplete. Potential aluminum toxicity appears tenable for some forest types (especially more sensitive hardwoods), but no immediate, high-risk situation is apparent.

Our assessment does not support the assumption that forest soils in the South will remain relatively unaffected by the deposition of S and N compounds. We anticipate that substantial portion (10%? to 30%?) of forest soils in the region will show major changes in soil chemistry within 50 yr. However, management effects may match or exceed the effects of acidic deposition, and the implications for forest health, productivity, and management should be examined together.

Recommended Research Directions

Our simulations showed that the effects of acidic deposition on soil chemistry depend strongly on the interactions of cations on the exchange complex and in solution as well as on the rate at which weathering resupplies basic cations to the exchange complex.

—Additional laboratory studies are needed to verify the relationships depicting cation exchange (selectivity coefficients) over a wide range of soil conditions. Subsequently, the sensitivity of simulation models to this process needs to be evaluated. In addition, work is needed to identify whether cation exchange relationships remain constant within pedons over time or if changes in the quality and quantity of organic matter or changes in pH affect cation exchange equilibria. In-field investigations of soil solutions and exchange complexes (perhaps examining cation dynamics in response to manipulations of ionic strength) are needed to verify the results of laboratory studies.

—Additional research is needed to evaluate the potential rates of mineral weathering in a range of soils in the region. We found no estimates of weathering rates for the Piedmont or Coastal Plain; this is perhaps the most important information gap in our assessment. The Integrated Forest Study (EPRI/ORNL) will provide information on a few additional sites, but a regional assessment will require information for representatives soils around the region.

—Once these two projects are complete, a regional survey to collect appropriate data and to run new simulations would lead to a substantial improvement relative to our study.

The ecosystem implications of acidification need to be clarified with experimentation on forest ecosystems. The supply rates of nutrient cations that may limit forest production should be evaluated with an extensive series of fertilization trials. Such trials should include not only applications of nutrient cations, but also common management treatments, such as nitrogen and phosphorus fertilization. Experimentation with aluminum concentrations in soil solution may not be as straightforward, but possible approaches include plot and whole ecosystem manipulation by liming (with sodium compounds to avoid confounding direct nutrient effects, and also with nutrient cation bases) and acid treatment. Although such manipulations may not represent likely changes in soil chemistry over a period of decades, some simple experimentation may serve to identify the degree to which nutrient cation supplies and aluminum concentrations warrant concern and more detailed research.

Additional research on sulfur retention is needed if more precise evaluations are desired to determine which soils may be expected to retain sulfate. However, available evidence indicates that many soils are already S saturated (if deposition estimates are accurate) relative to current deposition rates, so we expect additional research would be useful only in identifying classes of soils with similar sulfur status. We feel that enough information is available to conclude that many soils should not be expected to retain sulfate.

Over the longer term, the only clear verification of expected acidification trends will come from monitoring of field sites. Features of a monitoring network should includes:

—Sites expected to be sensitive and sites expected to be insensitive.

—Untreated sites and sites where minimal, annual additions of nutrients (and perhaps other treatments, such as liming) help screen out natural stand dynamics.

—Ranges of ages of stands within homogenous sites to help screen out stand age effects.

—Intensive monitoring sites (multiple samplings each year), perhaps maintained by government agencies or universities. Parameters monitored would include: ecosystem nutrient cycling, stand growth, species dynamics, soil microbiology, soil chemistry (perhaps modeled after the Integrated Forest Study [EPRI/ORNL]).

—Extensive monitoring sites with minimal sampling (once a year or decade), perhaps widely distributed on industrial lands and government-owned lands. Monitoring might include: periodic sampling of soils for determination of chemistry (pH, cation selectivity coefficients, ANC, BNC, acid strength, etc.) and stand growth.

Based on currently available information, we conclude that it is not safe to assume that acidic deposition will not lead to acidification of a substantial portion of the forest soils in the South. We are not ready to conclude that the converse is true (that acidic deposition will lead to acidification), and we believe that research along the lines outlined above will help narrow the degree of uncertainty. Most of the suggested research is also justified for gaining insights into general forest productivity and management and would be valuable in the absence of concerns over possible acidification.

References

Adams, D.F., S. Farwell, M. Pack, and E. Robinson. 1980. Estimates of natural sulfur source strengths. pp. 34–35 in *Atmospheric Sulfur Deposition: Environmental Impact and Health Effects*. (D.S. Shriner, C.R. Richmond, and S.E. Lindberg, eds.). Ann Arbor Science, MI.

Allen, H.L. 1987. Forest fertilizers. *Journal of Forestry* 85: 37–46.

Anderson, M. 1987. The effects of forest plantations on some lowland soils. I. A second sampling of nutrient stocks. *Forestry* 60: 69–85.

Arnon, D.I., and C.M. Johnson. 1942. Influence of hydrogen ion concentration on the growth of higher plants under controlled conditions. *Plant Physiology* 17: 525–539.

Baker, J.B., G.L. Switzer, and L.E. Nelson. 1974. Biomass production and nitrogen recovery after fertilization of young loblolly pines. *Soil Science Society of America Proceedings* 38: 958–961.

Berden, M., S.I. Nilsson, K. Rosen, and G. Tyler. 1987. Soil acidification—extent causes and consequences: an evaluaion literature information and current research. Report no. 3292, National Swedish Environment Protection Board, Solna. 164 pp.

Berger, K. 1962. Micronutrient deficiencies in the United States. *Journal of Agricultural and Food Chemistry* 10: 178–181.

Binkley, D. 1986. *Forest Nutrition Management*. Wiley, New York, 290 pp.

Binkley, D., and S. Hart. 1989. The components of nitrogen availability assessments in forest soils. *Advances in Soil Science,* in press.

Binkley, D., and D. Richter. 1987. Nutrient cycles and H^+ budgets of forest ecosystems. *Advances in Ecological Research* 16: 1–51.

Binkley, D., D. Valentine, C. Wells, and U. Valentine. 1989. Factors contributing to 20-year decline in soil pH in an old-field plantation of loblolly pine. *Biogeochemistry*, in press.

Binkley, D., P. Sollins, R. Bell, D. Sachs, C. Glassman, and D. Myrold. Biogeochemistry of adjacent conifer and alder/conifer ecosystems. In review.

Bloss, S., and D. Binkley. 1988. Effects of rooting by wild boars on nitrogen mineralization in high-elevation beech forests of the southern Appalachians. Submitted to *Canadian Journal of Forest Research*.

Bohn, H., B. McNeal, and G. O'Connor. 1985. *Soil Chemistry*. Wiley, New York. 340 pp.

Bolin, B., L. Granat, L. Ingelstom, M. Johannesson, E. Mattsson, S. Oden, H. Rodhe, and C.O. Tamm. 1972. Sweden's case study for the United Nations Conference on the Human Environment, 1972. Air pollution across national boundaries: the impact on the environment of sulfur in air and precipitation. Stockholm.

Boring, L., and W. Swank. 1984. The role of black locust (*Robinia pseudoacacia*) in forest succession. *Journal of Ecology* 72: 749–766.

Bormann, F.H. 1985. Air pollution stress on forests: An ecosystem perspective. *BioScience* 35: 434–441.

Bowden, R.D., C.T. GeBalle, and W. Breck Bowden. 1987. Foliar uptake of ^{15}N-labelled cloud/fogwater by red spruce (*Picea rubens*) seedlings. *Bulletin of the Ecological Society of America* 68: 267

Brand, D., P. Kehoe, and M. Connors. 1986. Coniferous afforestation leads to soil acidification in central Ontario. *Canadian Journal of Forest Research* 16: 1389–1291.

Brown, D.J.A. 1983. Effect of calcium and aluminum concentrations on the survival of brown trout (*Salmo tratta*) at low pH. *Bulletin of Environmental Contamination and Toxicology* 30: 582–583.

Bruck, R.I., and W.P. Robarge. 1987. Three-year survey of boreal montane forest decline in the southern Appalachian Mountains. pp. A34–A49 in NAPAP Terrestrial Effects Task Group IV. Peer Review, March 8–13, 1987, Atlanta, GA. Cited by permission of R.I. Bruck.

Bubenick, D.V. 1984. *Acid Rain Information Book*. Noyes Publications, Park Ridge, NJ.

Buol, S.W. (ed.) 1983. Soils of the southern states and Puerto Rico. Southern Cooperative Series Bulletin no. 174.

Buol, S.W., F.D. Hole, and R.J. McCracken. 1980. *Soil Genesis and Classification*. Iowa State University Press, Ames.

Chao, T.T., M.E. Harward, and S.C. Fang. 1964. Iron or aluminum coatings in relation to sulfate adsorption characteristics of soils. *Soil Science Society of America Proceedings* 28: 632–635

Chapin, F.S.H., K. Van Cleve, and P. Vitousek. 1986. The nature of nutrient limitation in plant communities. *American Naturalist* 127: 48–58

Clayton, J.L. 1987. Base transport from forested watersheds—weathering or exchange origin? *Agronomy Abstracts* 1987: 253.

Cleaves, E.T., D.W. Fisher. O.P. Bricker. 1974. Chemical weathering of serpentinite in the eastern Piedmont of Maryland. *Geological Society of America Bulletin*. 85: 437–444.

Cleaves, E.T., A.E. Godfrey, and O.P. Bricker. 1970. Geochemical balance of a small watershed and its geomorphic implications. *Geological Society of America Bulletin* 81: 3015–3032.

Cline, M., R.J. Stephens, and D.H. Marx. 1987. Influence of atmospherically deposited nitrogen on mycorrhiza: a critical literature review. SCFRC, Raleigh, NC.

Colquhoun, J., W. Kretser, and M. Pfeiffer. 1984. Acidity status update of lakes and streams in New York State. WMP–83. N.Y. State Department of Environmental Conservation, Albany.

Comerford, N.B., N.I. Lamson, and A.L. Leaf. 1980. Measurement and interpretation of growth responses of *Pinus resinosa* Ait. to K-fertilization. *Forest Ecology and Management* 2: 253–267.

Comerford, N.B., R.F. Fisher, and W.L. Pritchett. 1983. Advances in forest fertilization on the southeastern Coastal Plain. pp. 370–378 in *IUFRO Symposium on Forest Site and Continuous Productivity* (R. Ballard and S. Gessel, eds.). USDA Forest Service Pacific Northwest Forest and Range Experiment Station General Technical Report PNW–163. Portland, OR.

Cosby, B.J., G.M. Hornberger, and J.N. Galloway. 1985a. Modeling the effects of acid deposition: assessment of a lumped parameter model of soil water and streamwater chemistry. *Water Rescources Research* 21: 51–63.

Cosby, B., G. Hornberger, J. Galloway, and R. Wright. 1985b. Timescales of catchment acidification. *Environmental Science and Technology* 19: 1144–1149.

Cosby, B.J., G.M. Hornberger, R.F. Wright, and J.N. Galloway. 1986a. Modeling the effects of acid deposition: control of long-term sulfate dynamics. *Water Resources Research* 22: 1283–1291.

Cosby, B.J., R.F. Wright, G.M. Hornberger, and J.N. Galloway. 1985c. Modeling the effects of acid desposition: estimation of long-term water-quality responses in a small forested catchment. *Water Resources Research* 21: 1591–1601.

Cosby, B.J., P.G. Whitehead, and R. Neale. 1986b. A preliminary model of long-term changes in stream acidity in southwestern Scotland. *Journal of Hydrology* 84: 381–401.

Cowling, E. 1982. An historical resume of progress in scientific and public understanding of acid precipitation and its biological consequences. pp. 43–83 in *Acid Precipitation: Effects on Ecological Systems* (F.M. D'itri, ed.). Ann Arbor Science, MI.

Cronan, C.S. 1984. Biogeochemical responses of forest canopies to acid precipitation. pp. 65–79 in *Direct and Indirect Effects of Acidic Deposition on Vegetation* (R. Linthurst, ed.). Ann Arbor Science, MI.

Cronan, C.S. 1985. Biogeochemical influences of vegetation and soils in the ILWAS watersheds. *Water, Air, and Soil Pollution* 26: 355–372.

Cronan, C.S., and R.A. Goldstein. 1987. ALBIOS: an interregional analysis of aluminum biogeochemistry in forested watershed. *Agronomy Abstracts* 1987: 254.

Cronan, C.S., W.J. Walker, and P.R. Bloom. 1986. Predicting aqueous aluminum concentrations in natural waters. *Nature* 324: 140–143.

Daniels, R.B. 1987. Soil erosion and degradation in the southern Piedomont of the USA. pp. 407–428 in *Land Transformation in Agriculture* (M. Wolman and F. Fournier, eds.). Wiley, New York.

Daniels R.B., H.J. Kleiss, S.W. Buol, H.J. Byrd, and J.A. Phillips. 1984. *Soil systems of North Carolina*. North Carolina Agricultural Research Service Bulletin no. 467. 77 pp.

David, M.B., and C.T. Driscoll. 1984. Aluminum speciation and equilibria in soil solutions of a Haplorthod in the Adirondack Mountains, New York. *Geoderma* 33: 297.

David, M.B., M.J. Mitchell, and S.C. Schindler. 1984. Dynamics of organic and inorganic sulfur constituents in hardwood forest soils. pp. 221–245 in *Forest Soils and treatment Impacts* (E.L. Stone, ed.). Proceedings of the Sixth North America Forest Soils Conference, University of Tennessee, Knoxville.

Davidson, E., and W. Swank. 1986. Environmental parameters regulating gaseous nitrogen losses from two forested ecosystems via nitrification and denitrification. *Applied and Environmental Microbiology* 52: 1287–1292.

Davis, J.A., and J.O. Leckie. 1980. Surface ionization and complexation at the oxide/water interface. 3. Adsorption of anions. *Journal of Colloidal Interface Science* 74: 32–43.

Dickson, W. 1978. Some effects of acidification on Swedish lakes. *Verhandlungen Internat. Verein. Limnol.* 20: 851–856.

DiStefano, J. 1984. Nitrogen mineralization and non-symbiotic nitrogen fixation in an age sequence of slash pine plantations. Ph.D. dissertation, University of Florida, Gainesville. 219 pp.

Dochinger, L.S., and T.A. Seliga. 1976. *Proceedings First International Symposium on Acid Precipitation and the Forest Ecosystem*. USDA Forest Service General Technical Report no. 23. Upper Darby, PA.

Driscoll, C.T., J.P. Baker, J.J. Bisogni, and C.L. Schofield. 1980. Effect of aluminum speciation on fish in dilute acidified waters. *Nature* 284: 161–164.

Driscoll, C.T., N. van Breemen, and J. Mulder. 1985. Aluminum chemistry in a forested Spodosol. *Soil Science Society of America Journal* 49: 437–444

Driscoll, C.T., and W.D. Schecher, 1988. Aluminum in the environment. In Metal ions in biological systems. Marcel Dekker, in press.
Falkengren-Grerup, U. 1987. Long-term changes in pH of forest soils in southern Sweden. *Environmental Pollution* 43: 79–90.
Fenneman, N.M. 1938. *Physiography of the Eastern U.S.* McGraw-Hill, New York. 714 pp.
Firestone, M. 1987. The role of nitrate in the internal N-dynamics of a grassland and forest soil. *Agronomy Abstracts* 1987: 182.
Fisher, R.F. 1981. Soils interpretations for silviculture in the southern Coastal Plain. pp. 323–330 in Proceedings First Biennial Southern Silviculture Research Conference (J. Barnett, ed.). USDA Forest Service Gen. Tech. Rep. SO-34.
Fisher, R.F., and W.S. Garbett. 1980. Response of semi-mature slash and loblolly pine planations to fertilization with nitrogen and phosphorus. *Soil Science Society of America Journal* 44: 850–854.
Fölster, H. 1985. Proton consumption rates in holocene and present-day weathering of acid forest soils. pp. 197–209 in *The Chemistry of Weathering* (J. Drever, ed.). D. Reidel, Boston.
Fox, T.R., J.A. Burger, and R.E. Kreh. 1986. Effects of site preparation on nitrogen dynamics in the Southern Piedmont. *Forest Ecology and Management* 14: 241–256.
Foy, C.D. 1974. Effects of aluminum on plant growth. pp. 565–600 in *The Plant Root and Its Environment* (E.W. Carson, ed.). University Press of Virginia, Charlottesville.
Friedland, A.J., G.J. Hawley, and R.A. Gregory. 1985. Investigations of nitrogen as a possible contributor to red spruce (*Picea rubens* Sarg.) decline. pp. 95–106 in *Air Pollutant Effects on Forest Ecosystems*. Proceedings, May 8–9, St. Paul, MN. Acid Rain Foundation, St. Paul.
Friedland, A.J., G.J. Hawley, and R.A. Gregory. 1988. Red spruce (*Picea rubens Sarg.*) foliar chemistry in northern Vermont and New York, USA. *Plant and Soil* 105: 189–193.
Fuller, R.D., M.B. David, and C.T. Driscoll. 1985. Sulfate adsorption relationships in forested Spodosols of the northeastern U.S. *Soil Science Society of America Journal* 49: 1034–1040
Fuller, R.D., C.T. Driscoll, G.B. Lawrence, and S.C. Nodvin. 1987. Processes regulating sulfate flux after whole-tree harvesting. *Nature* 325: 707–710.
Galloway, J.N., G.E. Likens, and M.E. Hawley. 1984. Acid precipitation: natural versus anthropogenic components. *Science* 226: 829–831.
Gilliam, F. 1984. Effects of fire on components of nutrient dynamics in a lower coastal plain watershed ecosystem. Ph.D. dissertation, Duke University, Durham, NC.
Goldstein, R.A., C.W. Chen, and S.A. Gherini. 1985. Integrated lake-watershed acidification study: summary. *Water, Air, and Soil Pollution* 26: 327–337.
Gorham, E. 1955. On the acidity and salinity of rain. *Geochimica Cosmochimica Acta* 7: 231–239.
Grant, D., and D. Binkley. 1987. Rates of free-living nitrogen fixation in some Piedmont forest types. *Forest Science* 33: 541–548.
Grobstein, C. December 1983. Should imperfect data be used to guide public policy? *Science '83*, p. 18.
Hart, S.C., D. Binkley, and R. Campell. 1986. Predicting loblolly pine current growth and growth response to fertilization. *Soil Science Society of America Journal* 50: 230–233.
Harward, M.E., and H.M. Reisenauer. 1966. Reactions and movement of inorganic soil sulfur. *Soil Science* 101: 326–335.
Hauhs, M., and R.F. Wright. 1986. Regional pattern of acid deposition and forest decline along a cross section through Europe. *Water, Air, and Soil Pollution* 31: 463–474.
Hendrey, G.R., J.N. Galloway, S.A. Norton, C.L. Schofield, P.W. Shaffer, and D.A. Burns. 1980. *Geological and Hydrochemical Sensitivity of the Eastern United States to Acid Precipitation*. EPA 600/3-80-024. USEPA, Corvallis, OR. 100 pp.

Huag, A. 1984. Molecular aspects of aluminum toxicity. *CRC Critical Reviews in Plant Science* 1: 345–373.
Huang, C.P. 1975. Adsorption of phosphate on the hydrous $-Al_2O_3$ electrolyte interface. *Journal of Colloidal Interface Science* 53: 178–186.
Husar, R.B. 1986. Emissions of sulfur dioxide and nitrogen oxides and trends for eastern North America. pp. 48–92 in *Acid Deposition Long-Term Trends* (J. Gibson, ed.). National Academy Press, Washington, DC.
Hutchinson, T.C., L.Bozic, and G. Munoz-Vega. 1986. Responses of five species of conifer seedlings to aluminum stress. *Water, Air, and Soil Pollution* 31: 283–294.
Johnson, A., and A. Friedland. 1986. Recent and historic red spruce mortality: evidence of climate influence. *Water, Air and Soil Pollution* 30: 319–330.
Johnson, A., and S.B. McLaughlin. 1986. The nature and timing of the deterioration of red spruce in the northern Appalachian Mountains. pp. 200–230 in *Acid Deposition Long-Term Trends* (J. Gibson, ed.). National Academy Press, Washington, DC.
Johnson, D.W. 1984. Sulfur cycling in forests. *Biogeochemistry* 1: 29–43.
Johnson, D.W., and D.E. Todd. 1987. Nutrient export by leaching and whole-tree harvesting in a loblolly pine and mixed oak forest. *Plant and Soil* 102: 99–109.
Johnson, D.W., G.S. Henderson, and D.E. Todd. 1987. Changes in nutrient distribution in forests and soils of Walker Branch watershed, Tennessee, over an eleven-year period. *Biogeochemistry*, in press.
Johnson, D.W., D.W. Cole, H. Van Miegroet, and F.W. Horng. 1986. Factors affecting anion movement and retention in four forest soils. *Soil Science Society of America Journal* 50: 776–783.
Johnson, D.W., A.J. Friedland, H. Van Miegroet, R. Harrison, E. Miller, S. Lindberg, D. Cole, D. Schaefer, and D. Todd. 1988. Nutrient status of some contrasting high-elevation forests in the eastern and western United States. pp. Proceedings in U.S.–German Research Symposium, October 18–23, 1987, Burlington, VT.
Johnson, D.W., G.S. Henderson, D.D. Huff, S.E. Lindberg, D.D. Richter, D.S. Shriner, D.E. Todd, and J. Turner. 1982. Cycling of organic and inorganic sulfur in a chestnut oak forest. *Oecologia* 54: 141–148.
Johnson, D.W., J.W. Hornbeck, J.M. Kelly, W.T. Swank, and D.E. Todd. 1980. Regional patterns of soil sulfate accumulation—relevance to ecosystem sulfur budgets. pp. 507–520 in *Atmospheric Sulfur Deposition: Environmental Impact and Health Effects* (D.S. Shriner, C.R. Richmond, and S.E. Lindberg, eds.). Ann Arbor Science, MI.
Johnson, D.W., J.M. Kelly, W.T. Swank, D.W. Cole, J.W. Hornbeck, R.S. Pierce, and D. Van Lear. 1985a. *A comparative evaluation of the effects of acid precipitation, natural acid production, and harvesting on cation removal from forests*. Environmental Sciences Divison Publication no. 2508, Oak Ridge National Laboratory, TE.
Johnson, D.W., Richter, G.M. Lovett, and S.E. Lindberg. 1985b. The effects of atmospheric deposition on potassium, calcium and magnesium cycling in two deciduous forests. *Canadian Journal of Forest Research* 15: 773–782.
Johnson, N.M., C.T. Driscoll, J.S. Eaton, G.E. Likens, and W.H. McDowell. 1981. Acid rain, dissolved aluminum and chemical weathering at the Hubbard Brook Experimental Forest, New Hampshire. *Geochimica Cosmochimica Acta* 45: 1421–1437.
Johnston, A.E., Goulding, K.W.T., and P.R. Poulton. 1986. Soil acidification during more than 100 years under premanent grassland and woodland at Rothamstead. *Soil Use and Management* 2: 3–10.
Joslin, J.D. 1987. Predicting the response of red spruce seedlings to soil Al. *Agronomy Abstracts* 1987: 260.
Jung, G.E., and R. Werby. 1958. The concentration of chloride, sodium, potassium, calcium, and sulfate in rain water over the United States. *Journal of Meteorology* 15: 417–425.
Katz, B.G., O.P. Bricker, and M.M. Kennedy. 1985. Geochemical mass-balance relationships for selected ions in precipitation and stream water, Catoctin Mountains, *Maryland. American Journal of Science* 285: 931–962.

Kelley, J.K., J.D. Joslin, F.C. Thornton, M. Schaedle, and D. Raynal. 1987. A comparison of the response of red spruce seedlings to Al in soil and in solution culture. *Agronomy Abstracts* 1987: 260.
Kimmins, J.P., D. Binkley, L. Chatarpaul, and J. deCatanzaro. 1985. *Biogeochemistry of temperate forest ecosystems: literature on inventories and dynamics of biomass and nutrients*. Information Report PI–X–47 E/F, Canadian Forestry Service, Petawawa, ON 227 pp.
Klopatek, J.M., W.F. Harris, and R.J. Olson. 1980. A regional ecological assessment approach to atmospheric deposition: effects on soil systems. pp. 539–553 in *Atmospheric Sulfur Deposition: Environmental Impact and Health Effects* (D.S. Shriner, C.R. Richmond, and S.E. Lindberg, eds.). Ann Arbor Science, MI.
Levine, E.R., and E.J. Ciolkosz. 1988. Computer simulation of soil sensitivity to acid rain. *Soil Science Society of America Journal* 52: 209–215.
Likens, G.E., F.H. Bormann, N.M. Johnson, D.W. Fisher, and R.S. Pierce. 1970. Effects of forest cutting and herbicide treatment on nutrient budgets in the Hubbard Brook watershed-ecosystem. *Ecological Monographs* 40: 23–47.
Likens, G.E., R.F. Wright, J.N. Galloway, and T.J. Butler. 1979. Acid rain. *Scientific American* 241: 43–51.
Lindberg, S.E., G.M. Lovett, D.D. Richter, and D.W. Johnson. 1986. Atmospheric deposition and canopy interactions of major ions in a forest. *Science* 231: 141–145.
Lindsay, W. 1979. *Chemical Equilibria in Soils*. Wiley, New York. 449 pp.
MacCarthy, R., and C.B. Davey. 1976. Nutritional problems of *Pinus taeda* L. (loblolly pine) growing on pocosin soil. *Soil Science Society of America Journal* 40: 582–585.
McBride, M.B., and P.R. Bloom. 1977. Adsorption of aluminum by a smectite. II. An aluminum-calcium exchange model. *Soil Science Society of America Journal* 41: 1073–1077.
McFee, W.W. 1980. Sensitivity of soil regions to long-term acid precipitation. pp. 495–505 in *Atmospheric Sulfur Deposition: Environmental Impacts and Health Effects* (D.S. Shriner, C.R. Richmond, and S.E. Lindberg, eds.). Ann Arbor Science, MI.
McFee, W.W., E.J. Jokela, and N.B. Comerford. 1987. Slash and loblolly pine plantation response to micronutrients in the southeast. *Agronomy abstracts* 1987: 261–262.
McLaughlin, S.B., and H.S. Adams. 1987. Growth declines in red spruce. *Journal of Forestry* 85: 50–51.
Matzner, E., D. Murach, and H. Fortmann. 1986. Soil acidity and its relationship to root growth in declining forest stands in Germany. *Water, Air, and Soil Pollution* 31: 273–282.
Mead, D.J., and W.L. Pritchett. 1975. Fertilizer movement in a slash pine ecosystem. I. Uptake of N and P and N movement in the soil. *Plant and Soil* 43–451–465.
Mengel, K., and E.A. Kirkby. 1982. *Principles of Plant Nutrition*. International Potash Institute, Berne. 655 pp.
Meyer, J., R. Oren, and E.-D. Schulze. 1985. The effect of acid rain on forest tree roots: a review. pp. 16–30 in *Indirect Effects of Air Pollution on Forest Trees: Roots and Rhizosphere Interactions*. Proceedings of the Commission of the European Community Environmental Research Program.
Miller, R.E. 1979. Response of Douglas fir to foliar fertilization. pp. 62–68 in *Proceedings Forest Fertilization Conference* (S. Gessel, R. Kenady, and W. Atkinson, eds.). Institute of Forest Resources Contribution no. 40, University of Washington, Seattle.
Montagnini, F., B. Haines, and W. Swank. 1988. *Forest Ecology and Management*, in press.
Mooney, H.A., P.M. Vitousek, and P.A. Matson. 1987. Exchange of materials between terrestrial ecosystems and the atmosphere. *Science* 238: 926–932.
Morrison, I.K. 1984. Acid rain: a review of literature on acid deposition effects in forest ecosystems. *Forestry Abstracts* 45: 483–506.
Mulder, J., J. van Grinsven, and N. van Breemen. 1987. Impacts of acid atmospheric deposition on woodland soils in the Netherland. III. Aluminum chemistry. *Soil Science Society of America Journal* 51: 1640–1646.

Murray, G.E. 1961. *Geology of the Atlantic and Gulf Coastal Province of North America*. Harper's Geoscience Series, New York. 692 pp.

Myrold, D. 1987. Effects of acidic deposition on soil organisms. pp. 1–29 in *Acidic Deposition and Forest Soil Biology*. Technical Bulletin no. 527, National Council of the Paper Industry for Air and Stream Improvement, New York.

NADP/NTN. 1987. NADP/NTN annual data summary: percipitation chemistry in the United States. NREL, Colorado State University, Ft. Collins.

NAPAP. 1985. Annual report to the President and Congress.

NAPAP. 1986. Annual report to the President and Congress.

NCSFFC. 1980. Response of juvenile loblolly pine plantations to fertilization. NCSFFC Report no. 4. School of Forest Resources, North Carolina State University, Raleigh.

NCSFFC. 1982. Compilations of RW–no. 2 individual study reports (8-yr results). NCSFFC Report no. 13. School of Forest Resources, North Carolina State University, Raleigh.

Neal, C., P.G. Whitehead, R. Neale, and B.J. Cosby. 1986. Modeling the effects of acidic deposition and conifer afforestation on stream acidity in the British uplands. *Journal of Hydrology* 86: 15–26.

Neuman, K. 1986. Trends in public opinion on acid rain: a comprehensive review of existing data. *Water, Air, and Soil Pollution* 31: 1047–1059.

Nilsson, S.I. 1986. Critical deposition limits for forest soils. pp. 37–69 in *Critical Loads for Nitrogen and Sulfur: Report from a Nordic Working Group* (J. Nilsson, ed.). Nordisk Ministerrad Miljo Rapport 1986: 11.

Nilsson, S.I., and B. Bergkvist. 1983. Aluminum chemistry and acidification processes in a shallow podzol on the Swedish west coast. *Water, Ari, and Soil Pollution* 20: 311–329.

Olson, R.J., D.W. Johnson, and D.S. Shriner. 1982. *Regional assessment of potential sensitivity of soils in the eastern United States to acid precipitation*. Evnironmental Sciences Division Publication no. 1899, Oak Ridge National Laboratory, TN.

Paganelli, D., J. Seiler, and P. Feret. 1987. Root regeneration as in idicator of aluminum toxicity in loblolly pine. *Plant and Soil* 102: 115–118.

Paparozzi, E.T., and H.B. Tukey, Jr. 1984. Characterization of injury to birch and bean leaves by simulated acid precipitation. pp. 13–18 in *Direct and Indirect Effects of Acidic Deposition on Vegetation* (R. Linthurst ed.). Ann Arbor Science, MI.

Parker, D.R., L.W. Zelasny, and T.B. Kinraide. 1987. Chemical speciation and plant toxicity of aqueous aluminum in pp. 369–372. *Preprints of Papers Presented at the 194th National Meeting of the American Chemical Society*, vol. 27.

Parker, G.G. 1983. Throughfall and stemflow in the forest nutrient cycle. *Advances in Ecological Research* 13: 58–128.

Parker, G.G. 1987. Uptake and release of inorganic and organic ions by foliage: evaluation of dry deposition, pollutant damage, and forest health with throughfall studies. NCASI Technical Bulletin no. 532, New York.

Pastor, J., and W.M. Post. 1987. Influence of climate, soil mositure, and succession on forest carbon and nitrogen cycles. *Biogeochemistry* in press.

Pavan, M.A., F.T. Bingham, and P.F. Pralt. 1982. Toxicity of aluminum to coffee in Ultisols and Oxisols ammended with $CaCO_3$, $MgCO_3$ and $CaSO_4 \cdot 2H_2O$. *Soil Science Society of America Journal* 46: 1201–1207

Permar, T.A., and R.F. Fisher. 1983. Nitrogen fixation and accretion by wax myrtle (*Myrica cerifera*) in slash pine (*Pinus elliottii*) plantations. *Forest Ecology and Management* 5: 39–46.

Rastin, N., and B. Ulrich. 1985. Bodenchemische Standortskartierung zur Beurteilung des Stabilitatszustandes von Waldokosystemen in Hamburg. Ber. Forschungszent. Walskosysteme/Waldsterben, Göttingen Bd, 10: 1–91.

Reuss, J.O. 1980. Simulations of soil nutrient losses resulting from rainfall acidity. *Ecological Modeling* 11: 15–38.

Reuss, J.O. 1983. Implications of the calcium-aluminum exchange system for the effect of acid precipitation on soils. *Journal of Environmental Quality* 12: 591–595.

Reuss, J.O., and D.W. Johnson. 1985. Effect of soil processes on the acidification of water by acid deposition. *Journal of Environmental Quality* 14: 26–31.
Reuss, J.O., and D.W. Johnson. 1986. *Acid Deposition and the Acidification of Soils and Waters*. Springer-Verlag, New York. 119 pp.
Richter, D.D., P. Gromer, K. King, H. Sawin, and D. Wright. 1988. Effects of low ionic strength solutions on pH of acid forest soils. *Soil Science Society of America* 52: in press.
Riekrik, H., S.A. Jones, L.A. Morris, and D.A. Pratt. 1978. Hydrology and water quality of three small Lower Coastal Plain forested watersheds. *Proceedings of the Soil and Crop Science Society*, 38.
Robertson, G.P., P.M. Vitousek, P.A. Matson, and J.M. Tiedje. 1987. Denitrification in a clearcut lobolly pine (*Pinus taeda* L.) plantation in the southeastern U.S. *Plant and Soil* 97: 119–129.
Rochelle, B., and M. Church. 1987. Regional patterns of sulfur retention in watersheds of the Eastern U.S. *Water, Air, and Soil Pollution* 36: 61–73.
Rochelle, B., M. Church, and M. David. 1987. Sulfur retention in intensively studied sites in the U.S. and Canada. *Water, Air, and Soil Pollution* 33: 73–83.
Rost-Siebert, K. 1983. Aluminuium-Toxizitat und -Toleranz and Keimplfanzen von Fichte (*Picea abies Karst.*) und Buche (*Fagus silvatica* L.) *Allg. Forstz.* 38: 686–689.
Sasser, C.L., and D. Binkley. 1988. Nitrogen mineralization in high-elevation forests of the Appalachians. II. Patterns with stand development in fir waves. *Biogeochemistry* in press.
Satterson, K.A. 1985. Nitrogen availability, primary production, and nutrient cycling during secondary succession in North Carolina Piedmont forests. Ph.D. dissertation, University of North Carolina, Chapel Hill. 167 pp.
Schecher, W.D., and C.T. Driscoll. 1987. An evaluation of uncertainty in aluminum equilibrium calculations. *Water Resources Research* 23: 525–534.
Schutt, P., and E.B. Cowling. 1985. Waldsterben, a general decline of forests in Central Europe: symptoms, development, and possible causes. *Plant Diseases* 69: 548–558.
Schutt, P., W. Koch, H. Blaschke, K.J. Lang, E. Reigber, H.J. Schuck, and H. Summerer. 1983. *So stirbt der Wald*. BLV–Verlag, Munich. 127 pp.
Shriner, D.S. 1977. Effects of simulated rain acidified with sulfuric acid on host-parasite interactions. *Water, Air, and Soil Pollution* 8: 9–14.
Sheffield, R.M., N.D. Cost, W.A. Bechtold, and J.P. McClure. 1985. *Pine growth reductions in the southeast*. USDA Forest Service Resource Bulletin SE–83. Asheville, NC. 112 pp.
Singh, B.R. 1984. Sulfate sorption by acid forest soils. 3. Desorption of sulfate from adsorbed surfaces as a function of time, desorbing ion, pH, and amount of adsorption. *Soil Science* 138: 346–353.
Smil, V. 1985. *Carbon, Nitrogen, and Sulfur: Human Interference in Grand Biospheric Cycles*. Plenum Press, New York.
Smirnoff, N., P. Todd, and G.R. Stewart. 1984. The occurence of nitrate reduction in the leaves of woody plants. *Annals of Botany* 54: 363–374.
Smith, W.H. 1987. The atmosphere and the rhizosphere: linkages with potential significance for forest tree health. pp. 30–94 in *Acidic Deposition and Forest Soil Biology*. Technical Bulletin no. 527, National Council of the Paper Industry for Air and Stream Improvement, New York.
Smith, W.H., L.E. Nelson, and G.L. Switzer. 1971. Development of the shoot system of young loblolly pine. II. Dry matter and nitrogen accumulation. *Forest Science* 17: 55–62
Soil Survey Staff. 1975. Soil taxonomy. USDA Agriculture Handbook no. 436, 745 pp. Washington, DC.
Sposito, G. 1981. Cation exchange in soils: an historical and theoretical perspective. pp. 13–30 in *Chemistry in the Soil Environment*. ASA Special Publication no. 40, American Society of Agronomy, Madison, WI.
Stensland, G.J. 1984. Wet deposition network data with application to selected problems.

pp. 8–31 to 8–71 in *The Acid Deposition Phenomenon and its Effects: Critical Assessment Review Papers,* vol. 1. EPA–600/8–83–G16AF. Atmospheric Sciences, USEPA.

Stohr, D. (ed.) 1984. Waldbodenversauerung in Osterreich. Veranderungern der pH-Werte von Waldboden wahrend der letzten Dezennien. Osterreichischer Forstverein Forschungsinitiative gegen das Waldsterben, Institut fur Forstokologie, Vienna.

Strader, R., D. Binkley, and C. Wells. 1988. Nitrogen mineralization in high-elevation forests of the Appalachians. I. Regional patterns in spruce-fir forests. *Biogeochemistry* in press.

Stumm, W., and J.J. Morgan. 1981. *Aquatic Chemistry*. Wiley, NY.

Swank, W., and J. Waide. 1988. Characterization of baseline precipitation and stream chemistry and nutrient budgets for control watersheds. pp. 58–79 in *Forest Hydrology and Ecology at Coweeta* (W. Swank and D. Crossley, Jr., eds.). Springer-Verlag, New York.

Switzer, G.L., and L.E. Nelson. 1972. Nutrient accumulation and cycling in loblolly pine (*Pinus taeda* L.) plantation ecosystems: the first twenty years. *Soil Science Society of America Proceedings* 36: 143–147.

Tabatabai, M.A. 1985. Effect of acid rain on soils. *CRC Critical Reviews in Environmental Control* 15: 65–110.

Tabatabai, M.A. 1987. Physicochemical fate of sulfate in soils. *Journal of the American Physical Chemistry Association* 37: 34–38.

Tamm, C.O., and L. Hallbacken. 1986. Changes in soil pH over a 50-year period under different forest canopies in SW Sweden. *Water, Air, and Soil Pollution* 31: 337–341.

Thornbury, W.B. 1965. *Regional Geomorphology of the United States*. Wiley, New York. 609 pp.

Thornton, F.C., M. Schaedle, and D.J. Raynal. 1986a. Effect of aluminum on the growth of sugar maple in solution culture. *Canadian Journal of Forest Research* 16: 892–896.

Thornton, F.C., M. Schaedle, and D.J. Raynal. 1986b. Effects of aluminum on growth, development, and nutrient composition of honeylocust (*Gleditsia triacanthos* L.) seedlings. *Tree Physiology* 2: 307–316.

Thornton, F.C., M. Schaedle, and D.J. Raynal. 1987. Effects of aluminum on red spruce seedlings in solution culture. *Environmental and Experimental Botany* in press.

Tisdale, S.L., W.L. Nelson, and J.D. Beaton. 1985. *Soil Fertility and Fertilizers*. Macmillan, New York. 754 pp.

Trimble, S.W. 1974. Man-induced soil erosion in the southern Piedmont. Soil Conservation Soceity.

Turner, R.S., R.J. Olson, and C.C. Brandt. 1986. *Areas having soil characteristics that may indicate sensitivity to acidic deposition under alternative forest damage hypotheses*. Environmental Sciences Division Pulbication no. 2720, Oak Ridge National Laboratory, TN.

Turvey, N.D., and H.L. Allen. 1987. Site and cultural treatment effects and their interactions on four-year growth of loblolly pine. NCSFNC Report no. 19, School of Forest Resources, North Carolina State University, Raleigh. 31 pp.

Ulrich, B. 1983. Interaction of forest canopies with atmospheric constituents. pp. 33–45 in *Effects of Accumulation of Air Pollutants in Forest Ecosystems* (B. Ulrich and J. Pankrath, eds.). D. Reidel, Boston.

Ulrich, B. 1987. Stability, elasticity, and resilience of terrestrial ecosystems with respect to matter balance. pp. 11–49 in *Ecological Studies*, vol. 61 (E.-D. Schulze and H. Zwolfer, eds.). Springer-Verlag, Berlin.

Ulrich, B., E. Ahrens, and M. Ulrich. 1971. Soil chemical differences between beech and spruce sites—an example of the methods used. pp. 171–190 in *Integrated Experimental Ecology: Methods and Results of Ecosystem Research in the German Solling Project*. Springer-Verlag, New York.

Unger, F. 1836. Über den Einfluss des Bodens avf die Verteilung der Gewächse. Vienna.

USDA—Forest Service. 1969. A forest atlas of the South. Southeastern Forest Experiment Station. Asheville, NC. 27 pp.

USDA—Soil Conservation Service. 1966–1981. Soil Survey Investigations Reports nos. 2–37. Washington.
USDA—Soil Conservation Service. 1981. Land resource regions and major land resource areas of the United States. Agricultural Handbook no. 296. Washington.
USDA-Soil Conservation Service. 1984. 1982 National resource inventory. Washington.
U.S. Department of Commerce. 1983. Climatological atals. Environmental Data Service, Washington.
van Breemen, N., and J. Mulder. 1987. Atmospheric acid deposition: effects on the chemistry of forest soils. pp. 141–152 in *Acidification and Its Policy Implications* (T. Schneider, ed.). Elsevier, Amsterdam.
van Breemen, N., C.T. Driscoll, and J. Mulder. 1984. Acidic deposition and internal proton sources in acidification of soils and waters. *Nature* 307: 599–604.
van Breemen, N., J. Mulder, and C.T. Driscoll. 1983. Acidification and alkalinization of soils. *Plant and soil* 75: 283–308.
van Breemen, N., P.A Burrough, E.J. Velthorst, H.F. van Dobben, T. de Wit, T.B. Ridder, and H.F.R. Reijnders. 1982. Soil acidification from atmospheric ammonium sulfate in forest canopy throughfall. *Nature* 299: 548–550.
Van Lear, D.H., W.T. Swank, J.E. Douglass, and J.B. Waide. 1983. Forest management practices and the nutrient status of a lobolly pine plantation. pp. 252–258. in *IUFRO Symposium on Forest Site and Continuous Productivity* (R. Ballard and S. Gessel, eds.). USDA Forest Service General Technical Report PNW–163, Portland, OR.
Van Miegroet, H., and D.W. Cole. 1984. The impact of nitrification on soil acidification and cation leaching in a red alder ecosystem. *Journal of Environmental Quality* 13: 586–590.
Velbel, M.A. 1985. Geochemical mass balances and weathering rates in forested watersheds of the Southern Blue Ridge. *American Journal of Science* 285: 904–930.
Vitousek, P.M., and P. Matson. 1985. Disturbance, nitrogen availability, and nitrogen losses in an intensively managed loblolly pine plantation. *Ecology* 66: 1360–1376.
Waide, J.B., and W.T. Swank. 1987. Patterns and trends in precipitation and stream chemistry at the Coweeta Hydrologic Laboratory. pp. 421–430 in Aquatic Effects Task Group VI Peer Review Summaries Vol. II. North Carolina State University Atmospheric Impacts Research Program, Raleigh.
Weller, D.E., W.T. Peterjohn, N.M. Goff, and D.L. Correll. 1986. Ion and acid budgets for a forested Atlantic Coastal Plain watershed and their implications for the impacts of acid deposition. pp. 392–421 in *Watershed Research Perspectives*. Smithsonian Institution Press, Washington.
Wells, C.G., A. Jones, and J. Craig. 1988. Denitrification in southern Appalachian spruce-fir forests. Manuscript in review.
Wells, C.G., D.M. Crutchfield, N.M. Berenyi, and C.B. Davey. 1973. Soil and foliar guidelines for phosphorus fertilization of loblolly pine. USDA Forest Service Research Note SE–110, Asheville, NC.
White, G., S. Feldman, and L. Zelazny. 1988. Rates of nutrient release by mineral weathering. NCASI Technical Bulletin no. 542, New York.
Wiklander, L. 1974. The acidification of soil by acid precipitation. *Grundforbaettring* 26: 155–164.
Wiklander, L. 1980. The sensitivity of soils to acid precipitation. pp. 553–567 in *Effects of Acid Precipitation on Terrestrial Ecosystems* (T.C. Hutchinson and M. Havas, eds.). Plenum Press, New York.
Wiklander, L., and A. Anderson. 1972. The replacing efficiency of hydrogen ion in relation to base saturation and pH. *Geoderma* 7: 159–165.
Winner, W.E., T.B. Leininger, and S.B. McLaughlin. 1986. Forest responses to deposition of air-borne chemicals. pp. 22–44 in *Atmospheric Deposition and Forest Productivity*. Proceedings, Fourth Regional Technical Conference, Appalachian Society of American Foresters, January 29–31, 1986, Raleigh, NC.
Witter, J.A., and I.R. Ragenovich. 1986. Regeneration of Fraser fir at Mt. Mitchell,

North Carolina, after depredations by the balsam woolly adelgid. *Forest Science* 32: 585–594.
Wolt, J. 1987. Effects of acidic deposition on the chemical form and bioavailability of soil aluminum and manganese. Technical Bulletin no. 518, National Council of the Paper Industry for Air and Stream Improvement, New York.
Woodman, J. 1987. Pollution-induced injury in North American forests: facts and suspicions. *Tree Physiology* 3: 1–15.
Wright, R.F., B.J. Cosby, G.M. Hornberger, and J.N. Galloway. 1986. Comparison of paleolimnological with MAGIC model reconstructions of water acidification. *Water, Air, and Soil Pollution* 30: 367–380.
Zedaker, S., D.M. Hyink, and D.W. Smith. 1986. Growth declines in red spruce: are they anthropogenic or natural? *Journal of Forestry* 85: 34–36.
Zöttl, H.W., and R.F. Hüttl. 1986. Nutrient supply and forest decline in Southwest-Germany. *Water, Air and Soil Pollution* 31: 449–462.

Index

MIX
Papier aus verantwortungsvollen Quellen
Paper from responsible sources
FSC® C105338

If you have any concerns about our products,
you can contact us on
ProductSafety@springernature.com

In case Publisher is established outside the EU,
the EU authorized representative is:
Springer Nature Customer Service Center GmbH
Europaplatz 3, 69115 Heidelberg, Germany

Printed by Libri Plureos GmbH
in Hamburg, Germany